ESSAYS IN THE
PHILOSOPHY OF MATHEMATICS

Essays in the
Philosophy of Mathematics

by

R. L. GOODSTEIN, Ph.D., D.Lit.

Professor of Mathematics in the
University of Leicester

LEICESTER UNIVERSITY PRESS

1965

Printed by T. and A. Constable Ltd.
for Leicester University Press

PREFACE

THE essays in this volume discuss the nature of mathematics, the evolution of its concepts, the source of its truths, and the manner of its application to the real world. One of the essays, on logical paradoxes, has not been published before; the remainder have appeared in various journals over the past twenty-five years. The order in which they have been arranged is not the chronological one, but rather one which I think reflects the development of the central ideas. Attempts to clarify a concept seldom follow a single path and in the essays I have returned again and again to the same themes striving to say more exactly what I believed I saw clearly; the act of committing a thought to paper is intended to set it free, but all too often alas what the pen leaves is only the corpse of an idea.

Several of the essays were written to be read at gatherings of mathematicians and philosophers and I have not sought to change their style in preparing them for this collection; I have however made some alteration in their content, generally by deletion rather than addition. In the bibliographical notes which follow I have indicated both the occasion on which the lecture was given and the journal in which it was published.

R. L. GOODSTEIN

University of Leicester

June 1964

ERIC

IN MEMORIAM

BRAVEST OF SPIRITS

KEENEST OF INTELLECTS

FRAILEST OF BODIES

CONTENTS AND BIBLIOGRAPHICAL NOTES

I

PROOF BY REDUCTIO AD ABSURDUM

I F, like David Copperfield, you are on the road to Dover, for the first time and with no one to guide you, and you come to a point where the road forks, then of all your mathematical knowledge, only the method of *reductio ad absurdum* can help you. For if you choose the wrong turning, and find yourself in Folkestone, you may be quite sure you ought to have taken the other road.

Of course you may well consider that it would have been far better to have chosen the right road at the outset, and that is exactly how I feel about a *reductio ad absurdum* proof.

I am going to follow the usual practice in mathematics of calling any indirect proof, a proof by *reductio ad absurdum*, though there are indirect proofs, quite commonly met with in mathematics, which are not *reductions to a contradiction*; these are based on the formula

$$(not\text{-}q \text{ implies } not\text{-}p) \text{ implies } (p \text{ implies } q), \qquad (F)$$

that is, q follows from p if you deny p by denying q.

For instance, to prove

$$\text{if } \tfrac{1}{2}(1 + (-1)^n) \text{ is odd then } n \text{ is even}$$

it suffices, by formula (F), to observe that

$$\text{if } n \text{ is odd then } \tfrac{1}{2}(1 + (-1)^n) = 0, \text{ which is even;}$$

this is an indirect proof, but there is no apparent reduction to a contradiction.

The genuine *reductio ad absurdum* is a way of setting out an indirect proof which we have inherited from Euclid. In the last example, for instance, following Euclid, we should make two hypotheses:

(i) $\qquad\qquad\qquad \tfrac{1}{2}(1 + (-1)^n) \text{ is odd,}$
(ii) $\qquad\qquad\qquad\qquad\qquad n \text{ is odd,}$
and from (ii) deduce $\quad \tfrac{1}{2}(1 + (-1)^n) \text{ is even,}$

which contradicts the first hypothesis. The conclusion is that the hypotheses cannot both be true, i.e. the first hypothesis implies that the second is false. In modern logic we should call the hypotheses *axioms* and would say that a formal system postulating both the axioms (i) and (ii) contains a contradiction, and conclude that the axioms are incompatible.

Though different in form these two methods of proof, the first by means of formula (F), and the other by incompatible axioms, are logically equivalent, and I shall take the latter as the standard form.

Consider the following general scheme of a *reductio ad absurdum* proof:

axiom or proved proposition,	A
hypothesis,	*not-p*
theorem,	*not-p* implies *not-A*
conclusion,	*p.*

Evidently, the scheme is based on two presuppositions. In the first place we are assuming that A cannot be both true and false, for unless we reject the *contradiction* A and *not-*A, we have not made a case against *not-p*, and can draw no conclusion. We are also assuming that p is necessarily true or false, for if there were a third possibility over and above the truth and falsehood of p, then rejecting *not-p* would not lead to p because the third possibility would remain open.

These two assumptions, that a proposition cannot be both true and false, and that any proposition is *either* true *or* false, are called the laws of *contradiction* and *excluded middle* or *tertium non datur*. Unless these laws are established we are unable to justify the method of *reductio ad absurdum*, and in fact the whole weight of the attack which has been made on indirect proof in the past thirty years has fallen upon the law of excluded middle.

No one has yet succeeded in drawing a blank contradiction from an appeal to the *tertium non datur*. If that had happened there would be no controversy about the validity of indirect proof, because all mathematicians would reject the method. What the opponents of indirect proof have been concerned to show is that *reductio ad absurdum* leads to *absurd* rather than contradictory results. And, of course, what is, or is not, absurd is a matter of opinion.

The crux of the argument is the notion of existence in mathematics. By applying the law of contradiction and the *tertium non datur* to an existential proposition, the existence of a number with some specified property, or a function, or a set, is established by showing that non-existence leads to a contradiction, and this indirect method of proving existence is very widely used in mathematics in the so-called *pure existence theorems*. In elementary analysis, one of the most familiar of these pure existence theorems is the theorem which says that a non-decreasing bounded sequence converges to a unique limit; let us apply this theorem to a numerical sequence a_n, $n \geqslant 1$, defined as follows:

$a_n = 0$, if every even number from 2 to $2n$ is a sum of two primes,
$\quad = 1$ otherwise.

Since we can work out whether any given even number is, or is not, a sum of two primes (granted enough time, patience and paper), we can work out the value of a_n for any given n, and so the sequence is definite. In fact up to any r that has so far been tried, $2r$ is a sum of two primes, so that $a_r = 0$.

Since each term takes either the value zero or unity, the sequence is bounded, and it is non-decreasing because, if $a_n = 1$ for some n, then one of the even numbers from 2 to $2n$ is not a sum of two primes, and so $a_{n+1} = 1$. Our existence theorem tells us, therefore, that a_n has a unique limit. What is it? The sequence commences with a run of zeros; if every term is zero, then the limit is zero, but if one term takes the value unity, then so do the remainder, and the limit is unity. That is to say, if Goldbach's hypothesis that every even number is a sum of two primes, is true, then the limit is zero, and if the hypothesis is false then the limit is unity. Since we do not know whether Goldbach's hypothesis is true or not, we do not know which of the numbers zero or unity *is* the limit of the sequence, or for that matter, that either of them is a limit. On the one hand we have a proof that the limit exists, and on the other we are faced with the fact that we have no way of finding it. The sequence is perfectly definite, but the limit is not; and if we call the limit supplied by the existence theorem a *number l*, then the position of

l in the number series is indeterminate: for instance, we do not know if $l=0$ or if $l>0$. Now this is not a formal contradiction, for we have not proved that there is not a unique limit, but I suggest that, since it is impossible to find the limit, to say that we have proved there is a limit, is absurd.

Let me anticipate an obvious objection. It might be said: Surely we do know what the limit is; it is zero if Goldbach's hypothesis is true, and unity if the hypothesis is false. But I do not think this will do. In the first place the values of the terms of the sequence do not depend upon proving Goldbach's hypothesis, so why should the limit? Moreover, Goldbach's problem may be insoluble; but even if we do find the limit by proving Goldbach's hypothesis, it would not alter the fact that the existence theorem we started from did *not* determine the limit.

The existence theorem really adds nothing whatever to our knowledge of the sequence a_n.

We could express the conclusion we have reached by saying that propositions about sequences, for instance the proposition that a sequence s_n has a limit l, admit a third possibility over and above their truth and falsehood. Such propositions may be true, false or indeterminate (insoluble).

There is nothing mysterious about this idea of a third possibility, and it is quite easy to find 'illustrations' outside mathematics. Almost every day the newspapers tabulate the replies they receive to some question, in the form:

Yes .	.	.	.	37%
No .	.	.	.	60%
Don't know		.	.	3%

The law courts provide another example. English law applies the *tertium non datur*, for it holds a man innocent if he is not proved guilty, but Scottish law admits three verdicts, innocent, guilty and not-proven, so that in Scotland a refutation of guilt is not a proof of innocence. You can learn a lot about the foundations of mathematics by studying the law.

One of the fundamental differences between direct and indirect proof is that the former does not *presuppose* that the result we are seeking to prove makes sense. The direct proof supplies a meaning by joining the conclusion on to the general body of mathematical knowledge; whereas, in an indirect proof, we start by assuming that the conclusion is a significant proposition, one to which the law of excluded middle may be applied. This is shown rather strikingly in the following example.

Let us prove $\sqrt{(-1)} \leqslant 0$, by *reductio ad absurdum*.

If $\sqrt{(-1)} > 0$, then $\sqrt{(-1)} \times \sqrt{(-1)} > 0$, i.e. $-1 > 0$, which is false, and so $\sqrt{(-1)} \leqslant 0$. We all know this conclusion is nonsense, but what is wrong with the proof? Of course, we could proceed to prove also the contrary, that $\sqrt{(-1)} > 0$; for if $\sqrt{(-1)} \leqslant 0$ then $-\sqrt{(-1)} \geqslant 0$ and so $\{-\sqrt{(-1)}\}^2 \geqslant 0$, i.e. $-1 \geqslant 0$, which again is false. Do we conclude that mathematics is self-contradictory?

Just as the unsuccessful attempts to prove the parallel axiom culminated a century earlier in the discovery of non-Euclidean geometry, so the dispute over the validity of the *tertium non datur* led to many interesting and important discoveries in mathematics and symbolic logic.

The attack on *reductio ad absurdum* was started in 1913 by the Dutch mathematician L. E. J. Brouwer. At that time, Brouwer, who had already won a world-wide reputation by his work on topology, was one of the editors of the *Mathematische Annalen*, and he opened the attack by rejecting all papers offered to the *Annalen* which applied the *tertium non datur* to propositions the truth or falsehood of which could not be decided in a finite number of steps. The Editorial Board met this emergency by resigning—and then re-electing themselves, minus Brouwer. Incidentally, the Dutch Government so resented this slight on their leading mathematician that they founded a rival mathematical journal, with Brouwer in charge.

The defence of *reductio ad absurdum* was undertaken by Hilbert, who regarded Brouwer's criticisms as a personal affront. Hilbert planned to dispel all doubts of the validity of the indirect method, by PROVING that classical mathematics is free from contradiction, and contains no insoluble problems; this proof of course using only arguments accepted by everyone.

The formalist programme—as this ambitious plan was called —was abandoned in 1931 following the publication of Gödel's remarkable proof that every sufficiently rich mathematical system contains problems which are demonstrably insoluble. Gödel also established the impossibility of a proof of freedom from contradiction within the framework of classical mathematics. The question was re-opened, however, some five years later when Gentzen succeeded in proving that classical number theory is free from contradiction, but his proof transcended the limitations laid down in the formalist programme. Quite recently a system of mathematics has been constructed which is totally independent of the axioms of logic.

Another important consequence of the dispute was the discovery in 1920-21, independently by Emil Post in America and Lucasiewicz in Poland, of a multi-valued formal logic which rejects the *tertium non datur*. The discovery proved to be a fruitful one, and during the past ten years three-valued logics have been applied in such diverse fields as the foundations of mathematics, and quantum physics, in the latter case with striking success.

A logic is specified by its fundamental matrices (which is another way of saying it is an exercise in permutations and combinations). Writing T and F for true and false, respectively, T being considered the higher of the two values, the matrices for the classical two-valued logic are as follows:

p	not-p
T	F
F	T

p	q	p and q	p or q	p implies q	not-p	p or not-p
T	T	T	T	T	F	T
T	F	F	T	F	F	T
F	T	F	T	T	T	T
F	F	F	F	T	T	T

The first two columns simply exhibit the four ways of combining T and F. The third column defines 'p and q' as the function of p, q whose value is the *lower* of the values of p and q, and similarly the fourth column picks out the *higher* value to define 'p or q'. There are sixteen different ways of correlating a column of T's and F's with the first two columns, that is to say, sixteen functions of p, q; six of these are p, q, not-p, not-q, p and q, p or q, and it can easily be verified that the remaining ten are expressible in terms of these six (in fact, just the three functions *p, q, neither p nor q* suffice). For instance the function with values TFFT (in order) is

$$(p \ and \ q) \ or \ (not\text{-}p \ and \ not\text{-}q).$$

Note that *not-not-p* has the same values as *p*, and (*p implies q*) the same as (*not-p or q*). If you work out the values of the function *p or not-p* you find all T's; in other words (*p or not-p*) is true irrespective of whether *p* is true or false. It is not necessary for this proof of (*p or not-p*) that *p* be just an elementary proposition; the proof applies when *p* is any *finite* combination of elementary propositions, but it does *not* apply to an infinite set of propositions (like *every even number is a sum of two primes*). Thus the *tertium non datur* is demonstrable for propositions which have just two values. Accordingly any dispute about the *tertium non datur* is necessarily a dispute about the number of truth values of a proposition.

The corresponding matrices for a logic with three values true, indeterminate, false (in descending order) are

p	$not\text{-}p$
T	I
I	F
F	T

p	q	$p \ and \ q$	$p \ or \ q$
T	T	T	T
T	I	I	T
T	F	F	T
I	T	I	T
I	I	I	I
I	F	F	I
F	T	F	T
F	I	F	I
F	F	F	F
		lower value	higher value

Here *not-not-p* differs from *p*, but *not-not-not-p* has the same value as *p*; and instead of a *tertium non datur* we have a *quartum non datur*:

$$p \ or \ (not\text{-}p) \ or \ (not\text{-}not\text{-}p).$$

I should like to suggest now a rather different way of looking at the whole question of indirect proof. Instead of asking whether *reductio ad absurdum* is a valid method of proof, let us examine instead how much *information* such a proof gives us. We shall find that the answer varies from case to case, from no information at all at one end to all we could desire at the other. I shall try to illustrate this in the following examples.

1. As a first instance I shall take the well-known inequality:

if a, b are real and positive, $\frac{1}{2}(a+b) \geqslant \sqrt{(ab)}$.

By *reductio ad absurdum* we argue as follows: From

$$\tfrac{1}{2}(a+b) < \sqrt{(ab)}$$

we derive in turn

$$\{\tfrac{1}{2}(a+b)\}^2 < ab, \ a^2 - 2ab + b^2 < 0, \ (a-b)^2 < 0,$$

which contradicts the hypothesis that a and b are real. If we write this *reductio ad absurdum* proof *backwards*, simply replacing $<$ by $\geqslant$ we obtain a direct proof of the inequality, so that the direct and indirect proofs are simply journeys to and fro along the same road. As there can be no special merit in proceeding one way rather than the other, in this case direct and indirect proof are equally successful, and give the *same* information.

2. *The infinity of primes.* Here is the current *reductio ad absurdum* proof. Let N be the greatest prime. Since $N! + 1$ is not divisible by any of the numbers 2 to N, its least factor (above unity) must be a prime greater than N, which contradicts the hypothesis that N is the greatest prime. Compare this with a direct proof. Make no assumption about N and consider $N! + 1$; the least factor of this number is a prime greater than N. Hence if $N_0 = 2$ and N_{r+1} is the least factor (above unity) of $N_r! + 1$ for all r, then N_0, N_1, N_2, ... is an increasing infinite sequence of primes, and there is no greatest prime.

The direct and the indirect proof give the same information, namely the formula 'the least factor of $N! + 1$' for a prime greater than any N. My objection to the indirect proof in this case is on grounds of elegance. The indirect proof makes a quite superfluous assumption.

3. *The irrationality of $\sqrt{2}$.* I need not reproduce the details of the classical *reductio ad absurdum* proof. It will suffice to mention that from the assumption that there is a fraction p/q, where p, q are relatively prime, such that $(p/q)^2 = 2$, we deduce that p, q have a common factor 2 contrary to the hypothesis.

How much information is contained in this *reductio ad absurdum*? Does the proof tell us, for instance, by how much $\sqrt{2}$

differs from p/q? It does not; and if the proof proves $\sqrt{2}$ is irrational, then it proves, presumably, that in the expansion of $\sqrt{2}$ a digit cannot recur infinitely often. Well, how many times can a digit recur in this expansion? Again the proof is silent. In fact, the proof gives no information whatever other than that the assumption $(p/q)^2 = 2$ leads to a contradiction.

The surprising thing is that a direct proof gives all this information without adding one iota to the amount of work we have to do.

For considering odd and even values of p, q we see that $p^2 - 2q^2$ is a product of an odd number with a power of 2, and so

$$|p^2 - 2q^2| \geqslant 1,$$

whence
$$|p^2/q^2 - 2| \geqslant 1/q^2 \qquad \ldots \text{ (i)}$$

This is a direct proof that 2 is not the square of a rational number, which at the same time sets a lower bound to the difference. Let x_n be the expansion of $\sqrt{2}$ to n places of decimals, so that

$$x_n^2 < 2 < (x_n + 1/10^n)^2;$$

then from inequality (i) we readily derive

$$|x_n - p/q| \geqslant 1/6|pq|, \qquad \ldots \text{ (ii)}$$

provided that n exceeds the number of digits in q^2.

If, further, $x_n = 1 \cdot a_1 a_2 \ldots a_{n-1} \alpha \alpha \ldots \alpha a_{N+1} a_{N+2} \ldots$, and

$$x = 1 \cdot a_1 a_2 \ldots a_{n-1} \alpha = (9i + \alpha)/9 \cdot 10^{n-1} = p/q, \text{ say,}$$

where i, p, q are integers, then $|x - x_m| < 1/10^n$, for $m \geqslant N$; but
$$|x - x_m| \geqslant 1/6|p/q|q^2 =$$
$$1/6x \cdot 9^2 \cdot 10^{2n-2} > 1/9^3 \cdot 10^{2n-2} > 1/10^{2n+1}.$$

Hence $N < 2n + 1$, and so a recurrence in the expansion of $\sqrt{2}$ which commences at the nth digit terminates not later than the $(2n)$th.

For the sake of completeness I should like to give also a direct proof of the trivial theorem with which I opened this lecture. We have to prove that if $\frac{1}{2}(1 + (-1)^n)$ is odd, then n is even. Let $R(p)$ be the remainder when p is divided by 2; then, considering odd and even values of p,

$$R(p) \cdot R(\tfrac{1}{2}\{1 + (-1)^p\}) = 0, \text{ for all } p \qquad \ldots \text{ (i)}$$

B

Since $R(\frac{1}{2}\{1 + (-1)^n\}) = 1$, therefore $R(n) = 0$, and so n is even.

This of course is a proof by cases, but it may readily be transformed into a pure inductive proof as follows.

We define $\sigma(0) = 1$, $\sigma(p+1) = 0$; $R(0) = 0$, $R(p+1) = \sigma(R(p))$, then

$$R(p+2) = R(p), \qquad \ldots \text{ (ii)}$$

for $R(2) = \sigma(\sigma(R(0))) = \sigma(1) = 0 = R(0)$, so that (ii) holds for $p = 0$, and if (ii) holds for $p = k$, then

$$R(k+3) = \sigma(R(k+2)) = \sigma(R(k)) = R(k+1),$$

which completes the proof of equation (ii). Let $\phi(p)$ denote $R(\frac{1}{2}\{1 + (-1)^p\})$, then we have

$$R(0) = 0, \quad R(1) = 1, \quad R(p+2) = R(p)$$

and

$$\phi(0) = 1, \quad \phi(1) = 0, \quad \phi(p+2) = \phi(p).$$

Hence if

$$\delta(p) = |\phi(p)R(p) - \phi(p+1)R(p+1)|$$

then $\delta(0) = 0$, and $\delta(p+1) = \delta(p)$, so that $\delta(p) = 0$ for all p, that is, $\phi(p)R(p) = \phi(p+1)R(p+1)$; but $\phi(0)R(0) = 0$, and so $\phi(p)R(p) = 0$ for all p, which proves equation (i). It remains to show that if $R(n) = 0$ then n is even. If $W(p)$ is defined by the equations

$$W(0) = 0, \quad W(p+1) = W(p) + R(p),$$

then, observing that $R(p) + R(p+1) = 1$, a simple induction proves

$$p = 2W(p) + R(p);$$

hence if $R(n) = 0$ then $n = 2W(n)$, and so n is even.

One last observation. Let us apply a *reductio ad absurdum* argument to the unsolved Goldbach's problem.

It is evident that if we could find a number k so that every even number is a sum of two primes provided only that the first k even numbers are sums of two primes, then the problem is solved. We have only to test the first k even numbers to find out whether Goldbach's hypothesis is true or not.

Now to the proof that this number k exists. Consider the sequence a_n we introduced before:

$a_n = 0$ if $2r$ is a sum of two primes for all r from 1 to n,
 $= 1$, otherwise.

The sequence a_n is bounded and non-decreasing, and so convergent. Hence there is a k such that

$$a_n - a_k < 1/2 \text{ for all } n > k.$$

If for this k, $a_k = 0$, then all a_n are zero, and Goldbach's hypothesis is true; if $a_k = 1$ then the hypothesis is false. For a given k we can work out the value of a_k; I leave it to you just to test whether this a_k is 0 or 1. Of course, the same argument may be applied to any other unsolved problem in the theory of numbers —for instance, to Fermat's last problem.

How much information does this proof give you?

REFERENCES

Brouwer, L. E. J. 'Intuitionism and Formalism', *Bull. Amer. Math. Soc.*, 20, 1913.
Gentzen, G. 'Die Widersruchsfreiheit der reinen Zahlentheorie', *Math. Ann.*, 112, 1936.
Gödel, K. 'Über formal unentscheidbare Sätze der *Principia Mathematica* und verwandter Systeme. I', *Monatsheft f. Math. Phys.*, 38, 1931.
Goodstein, R. L. 'Function theory in an axiom-free equation calculus', *Proc. of the London Math. Soc.*, Ser. 2, 48, 1945.
——'Mathematical Systems', *Mind*, XLVIII, N.S., 189, 1939.
Heyting, A. 'Mathematische Grundlagenforshung, Intuitionismus, Bewistheorie', *Erg. d. Math.*, III, 4, Berlin, 1934.
Hilbert, D. 'Über das Unendliche', *Math. Ann.*, 95, 1926.
Lucasiewicz, J. 'Philosophische Bemerkungen zu mehrwertigen Systemen des Aussagenkalküls', *C. R. Soc. Sciences Varsovie*, 23, 1930.
Post, E. L. 'Introduction to a General Theory of Elementary Propositions', *Amer. Journ. Math.*, 43, 1921.
Reichenbach, H. *Philosophic Foundations of Quantum Mechanics*, University of California Press, 1946.

II

LOGICAL PARADOXES

I AM going to consider a variety of paradoxes, some in serious vein, and some not so serious, starting with two in the latter category.

The paradox of the barber

In a certain village the barber shaves all the men, and only those who do not shave themselves. Who shaves the barber? If the barber does not shave himself, then as one of those who do not shave themselves, he is shaved by the barber. And if he does shave himself then the barber does not shave him, yet it is the barber who shaves him.

Most people hearing the barber paradox for the first time soon come up with the suggestion that the barber is a woman (or a young boy), but we can readily block this loophole by declaring the barber to be a man, and more specifically a man who pays daily tribute to the blade. With this door closed the next suggestion will probably be that the barber is not himself one of the villagers, but this objection can again be met by sharpening the wording. The interesting question is: Why do we feel so driven to find a way out of the paradox? I think it is because we appear to be describing a real situation—the village barber really does shave those who do not shave themselves. I shall not pursue the task of sharpening the wording to leave no loophole whatever; instead I shall take it for granted that this can be done and see where the supposition takes us.

How does the situation described in the paradox differ from the real state of affairs in the village barber shop? Can we not say of any village that the village barber shaves all who are shaved and do not shave themselves? Of course we can, but when we say this we take it for granted that the totality we are talking about does not include the barber. Just as when we say that the teacher teaches all who attend the class. The paradox arises when we bring into question the class which the universal *all* comprehends. If this class does not include the barber then all is well—but if it does then we are in the paradox. It is because we are trapped by the word *all* that the tale of the barber at first seems reasonable and then astounds us by becoming contradictory. Yet we should not be at all surprised by a contradiction arising from a contradictory definition. Suppose for instance that we seek to define a class K by saying that K contains *a* and *b*, and contains *c* if and only if it does not contain *c*, i.e.

$$c \in K \longleftrightarrow c \notin K$$

We should at once conclude that K does not exist since the assumption that it exists leads to a contradiction. And what now of the barber paradox?—the paradox is nothing but a proof that there is no barber in the village. Everyone shaves himself. If the assumption that there is a barber leads to a contradiction, then this assumption is false, and there is no barber.

A paradox very similar to that of the barber is

The paradox of the reference book

A librarian noticed one day that some of the reference books in the library list themselves whereas others do not. He decided to pass the time producing a reference book listing all the reference books which do *not* refer to themselves. Just as he was on the point of completing the task he asked himself if he should include a reference to this new volume or not. If I do not list it, he said to himself, this will be a reference book which does not refer to itself, and so it should be listed; but if I do list it, this will not be one of the reference books I set out to list, and it should not be included. To make the matter definite let us suppose that the library contains 25 reference books A to Y, and that the new volume is Z. Z lists A if A does not list itself. Shall Z list itself or not, that is the question.

Let us again start by looking at the totality involved in the paradox. The librarian proposes to list all the reference books in the library which do not refer to themselves. This totality is clearly a function of time; must we not therefore impose a time limit? If we say for instance that Z contains all the relevant books in the library at noon today, then the paradox vanishes;

at 12.05 volume 26 is closed without a reference to itself being required. But suppose we refuse to accept a time limit? Then again the paradox vanishes because by definition Z can never be closed and the question of listing Z cannot arise. Z is a project, never a completed volume.

Both the paradox of the barber and that of the reference books are unsuccessful attempts to express Russell's paradox of classes in a popular form.

The Paradox of classes

This is the paradox of normal classes. A class is normal if it does not contain itself; for instance the class of persons in any room is not itself a person, and so this class is normal. But presumably the class of everything is a thing and so a member of itself, so that the class of everything is abnormal. Now consider the class N of all normal classes. Is N normal? If it is, then it does not contain itself—but as a normal class it is a member of N. If N is abnormal, then N contains itself; but the members of N are the normal classes. Thus N can neither be normal, nor abnormal. In symbols the definition of N is

$$x \in N \longleftrightarrow x \notin x$$

Taking N for x, we arrive at the contradiction

$$N \in N \longleftrightarrow N \notin N.$$

There are many ways in which we may seek to escape from the paradox. Our previous experience suggests that we ought to seek to show that N does not exist. Can we do this? Well, how is N defined? N is the class of all normal classes; thus to exclude N we must either deny that normality is a legitimate property, or we must ban the formation of universal classes. Russell took the first course and made normality illegitimate by means of a theory of types. According to this theory, objects are ranged in type hierarchies, with individuals at the lowest level, classes of individuals at the next level, and so on. The theory of types bans the formation of classes with members chosen from different levels. Thus the theory of types imposes a restriction on the membership relation. If $x \in y$, then y must belong to the next higher level than x; in particular $x \in x$ is ruled out, and so

normality and abnormality are forbidden concepts. Type theory
cuts out the paradox, but at a very high price in complexity in
the development of mathematics. It requires us, e.g., to suppose
that there is a number system in each type level, so that for
instance we have, a *two* of each type instead of a single number
two. Type theory is arbitrary and artificial. The class consisting
of Miss Jones and her IVth form seems a perfectly legitimate
class, but it is banned by type theory.

There are several alternative class theories. Zermelo banned
universal classes and allowed the formation only of subclasses
with a given property. A theory of Quine imposes a condition of
elementhood on objects which can be comprehended into a
class; the Russell paradox is thereby turned into a proof that a
certain object is not an element: The elementhood condition on
x is

$$(\exists y)(x \in y)$$

The comprehension operator 'the class of all x such that—' is
restricted to *elements* x. Thus let w be the class of all elements
which are not members of themselves

i.e. $\qquad x \in w \leftrightarrow (\exists y)(x \in y) \ \& \ x \notin x;$
then $\qquad w \in w \leftrightarrow (\exists y)(w \in y) \ \& \ w \notin w$

from which we readily deduce

$$\neg\,(\exists y)(w \in y)$$

that is, w is no element. Thus the Russell paradox is transformed
into a proof that w is not an element. Unfortunately the element-
hood condition alone does not suffice to protect class theory
from contradiction.

In another theory both classes and sets play a part; but sets
alone may be combined to form new classes: in such theories set
plays the part of collections and class that of the extension of a
predicate; i.e. to any predicate P corresponds a class K such
that

$$a \in K \leftrightarrow P(a)$$

but K cannot be the value of a set variable; only if there is a set
k with the same members as K can the objects with property P
form a collection.

The paradox of the surprise inspection

A paradox which has aroused considerable interest in the past few years is the paradox of the surprise inspection. A company commander announced one Sunday that there would be a surprise inspection at noon one day in the following week. His Lieutenant, a logician drafted into the armed forces, thought about the inspection and came to the conclusion that it could not take place. He argued like this:

If the inspection does not occur by noon on Saturday it will have to be on Sunday—and it will be no surprise. Accordingly, if it does not occur by noon on Friday, it will have to on Saturday, and again it will be no surprise. This drives us back to Thursday, for if it does not occur on Thursday, it will have to happen on Friday, and it will be no surprise; and so on right back to the beginning. Apparently the surprise inspection cannot take place and yet when it was announced on Sunday, we had no idea when it was to happen.

The paradox is not just a play on the word surprise. We are saying that an event is no surprise if we can deduce from the evidence before us when it will take place, before it actually happens. Thus if it is my invariable practice to visit Aunt Maud or Aunt Matilda each Sunday, and if I admit on Saturday that I am not visiting Aunt Matilda the next day, then my visit to Aunt Maud is no surprise; from my invariable practice, and my admission, Maud can deduce on Saturday that I shall visit her next day.

Let us now return to the original paradox and eliminate the superfluous trappings. Suppose that I say at noon: I will pay you a surprise visit at 1 o'clock or 2 o'clock today. When I say this I may not know myself when I shall pay the visit, and perhaps have to look up the trains. Yet arguing as the Lieutenant argued, if I do not come by 1 o'clock, you will know to expect me at 2, and my visit will be no surprise; hence I must arrive at 1 o'clock—and since you have deduced my arrival time, my visit at 1 o'clock will still be no surprise.

You may by now be content to accept the argument that a surprise visit on a finite number of alternative dates can be no surprise. Yet this seems to be quite contrary to experience. If someone visits us once a year regularly, but not necessarily on the same day, then he still surprises us when he does not announce his arrival day.

Let us think what may happen at 1 o'clock. If I actually arrive, can you say that you knew I would arrive at 1 o'clock? Clearly not, since you proved I would not come at all. You must greet my arrival with the words: how surprised I am to see you; I deduced from what you said that you would not come! Thus we surprise our friends when we do what we say we wont, not when we do what we say we will—or at anyrate we may surprise them.

Does this really resolve the paradox? Whether I actually turn up or not, has not my intention of paying a surprise visit been shown to be contradictory. If the paradox rests on the conflict with common sense, or common experience, then I think what I have said does dispose of it. Whether you reason as the Lieutenant did, or not, my coming at 1 o'clock is a surprise. If I believe someone cannot come (because he is in prison or because the trains are not running, or because I deduced from what he said that he was *not* coming) and if he does come, then his visit is a surprise, because I deduced from the evidence that he could not come.

I accept the Lieutenant's argument as a correct deduction that I will not come from my declared intention of coming unexpectedly at 1 or 2 o'clock. If I say that I will come at 1 o'clock and will be unexpected (by those to whom I am speaking) then I am in effect affirming, if p then not-p, where 'p' is 'I will come at one o'clock', and this is a refutation of p. In the same way, to say that I will come at 1 or 2 o'clock and my visit will be unexpected is, by the Lieutenant's argument a refutation of 'I will come at 1 or 2 o'clock'. This is surprising but it is no paradox.

The liar paradox

A paradox of great age and interest is the paradox of the liar. Suppose that I announce 'I am lying'. If I speak this truly then I am not lying though I say that I am; and if I speak falsely, then what I say is not the case, and so I am not lying. Thus I cannot say that I am lying without contradicting myself. Another, and perhaps clearer form of the same paradox is this:

The first sentence on page 1 of Zygmund's treatise on logic reads: **The first sentence on p. 1 of Zygmund's treatise on logic is false.** Call this sentence p. If p is true, then the first sentence in Z's logic is false, that is, p is false. And if p is false then the first sentence in Z's logic is true, i.e. p is true. Thus p is self-contradictory; p is an example of what is called self reference, for p appears to say something about its own meaning, namely that it is false.

Now it is quite a common thing for a sentence to say something about itself, that is about its representation. For instance the sentences

'This sentence is written in white chalk'
'This is an English sentence'
'This is written in Roman letters'

all say something about the form in which they are expressed. But can a sentence say something about its meaning? According to Wittgenstein the meaning of a sentence is the part which it plays in the whole language; how then can a sentence say something about its own role? The actor plays a part, he does not comment on his performance. Are we then to conclude that p, the first sentence in Z's logic, is no sentence? It is correctly formed, so on what grounds shall we reject it; how are such non-sentences to be recognised, or is it only by their contradictory fruits that we shall know them? And how can we be sure of any sentence that no one will derive a contradiction from it?

I like to think about this paradox in the setting of a yes-no questionnaire.

Question		Answer
1	Is $2 + 2 = 5$?	No
2	Is the answer to Question 1 No?	Yes
3	Is the answer to Question 3 No?	

Whether we now write Yes or No in the answer column we contradict ourselves. There is no doubt that Question 3 is reflexive since 'Question 3' is named in the question. Let us look more closely at the seemingly similar questions 2 and 3. What answer must we look at to reply to question 2? The answer to question 1. What answer must we look at to reply to question 3? The answer which we have yet to give to question 3, the answer not yet written in the empty box. Like those who answer the Gallup polls, we should like to register a don't know to question 3.

Now in question 2, the term 'question 1' which appears is a proper name, and it may be replaced by the sentence which it names. Names are merely a dispensable convenience. But what of the name 'Question 3' in question 3 itself? What sentence does this name? Replace the name by the question and we obtain the sentence 'What is the answer to the question "What is the answer to question number 3" '. And of course the attempt to eliminate the name leads to an infinite regress and the source of the paradox is seen to be an improper use of a name. It is as if I sought to define a function by saying,

for any n $\quad\quad f(n-1)=f(n)+n$
so that

$$f(0)=f(1)+1=f(2)+2+1=f(3)+3+2+1$$

and so on, and we never reach the value of $f(0)$. We seem to be defining a function, but in fact we are not. In the same way 'What is the answer to question 3' looks like a question but it is not. The paradox about the first sentence in Z's book is in no way different; the pseudo-description 'the first sentence in Z's book', cannot be eliminated.

It is often said that, whatever the fate of earlier attempts, Gödel certainly showed that self-reference is possible by means of arithmetic. Let us suppose that every English sentence is given a number in some way. E.g. each letter in the sentence may be replaced by as many ones as there are in its alphabet ordinal, followed by a zero. Then the whole sentence will be replaced by a run of 0's and 1's which we may take to be its number in the scale of 2. E.g. 'Back soon' becomes
11010110011111111110111111111111111111111101111111111111111111110111111111111111111110111111111110
It does not matter what method we use as long as the sentence determines its number uniquely, and the number determines the sentence. Suppose now that 99 turns out to be the number of the sentence

P: sentence ninety-nine is false.

Have we not now achieved genuine self-reference? I think not, for the position is exactly the same as in our yes-no game. If we seek to replace the description 'sentence ninety-nine' by P itself, the resulting sentence still contains a description, and we are back again in an infinite regress. The fact that sentences are now numbered by a rule, instead of by an arbitrary choice, makes no difference at all.

This example is however an oversimplification of what is meant by Gödel numbering. One can give a clearer picture of the method in terms of a code. Let us imagine that we have a code in which English sentences are encoded by certain other English sentences. Thus the code for 'the cat sat on the mat' might be 'this little pig went to market'. Let us suppose further that the code sentences are numbered off, and let us denote by c the third sentence in the code. Finally let us imagine that by accident or design c is the code for

the third code sentence is false.

Is not c now a true example of self-reference? To understand c we uncode it and find that c says

the third code sentence is false

and of course the third sentence in question is c itself. Thus c asserts its own falsity. Yet is this really so? When we say that the third code sentence is false we deny c itself, not what c encodes. Thus 'c' might be 'some cats are dogs', and to say that the third sentence is false is to say that no cats are dogs. If I send someone who knows the code the message c, and if he knows that the message is in code, he will correctly read it to say

the third code sentence is false;

he will then look up the third code sentence, and find that it is c itself, and he will then conclude that no cats are dogs. The code has been both *used* and *mentioned*, and there is no self reference. A sentence in one language may very well say something about the meaning of that sentence in another language.

Of course you may object and say: when I assert that the third code sentence is false, I mean that which it encodes is false. Let us then suppose that there exists a code in which some symbol ξ is the code word for 'the sentences encoded by ξ is false'. Writing 'ξ' for the sentence encoded by ξ, we are affirming

$$\text{'}\xi\text{'} = \text{'}\xi\text{'} \text{ is false}$$

which is false. But this is no paradox, it is simply a proof that the code in question does not exist.

And if you object again and insist that you are defining the code by declaring that ξ is the code word for 'ξ' is false, then you are back in an infinite regress, because you have failed to say what ξ stands for.

The paradox of Achilles and the tortoise

I have saved up for the end my favourite paradox, Achilles and the Tortoise. Let Achilles run ten times as fast as the tortoise, and let him give the tortoise 90 yards start in a hundred-yards race. When the tortoise has covered 10 yards Achilles has run 100 yards and so they arrive simultaneously at the winning post. It sounds all right, but is it? When the tortoise is half way home, Achilles halves the lead, when the tortoise has only a quarter of the run left, Achilles has reduced the gap lead to a quarter of its amount, and so. Each time the tortoise covers a half of the distance left, Achilles halves the tortoise's lead, but nevertheless even though he halves the lead, he remains behind; now although you may make the lead as small as you please by successive halving, you can never make it actually zero, and so Achilles never catches the tortoise.

For centuries mathematicians misunderstood the paradox and thought that it merely showed its poser Zeno was ignorant of the fact that an infinite series may have a finite sum. Thus they all argued that if the tortoise moved at, for example, a yard a second then he is half way to the winning post in 5 seconds, half way further on, after a further $\frac{5}{2}$ second, half way further on again, after a further $\frac{5}{4}$ second, and so on; the sum of these time intervals is

$$5(1 + \tfrac{1}{2} + \tfrac{1}{4} + \ldots) = 5 \cdot 2 = 10$$

agreeing with the fact that in 10 seconds Achilles catches the tortoise. Now what could be more absurd than to suppose that Zeno did not recognise that

$$\tfrac{1}{2} + \tfrac{1}{4} + \tfrac{1}{8} + \ldots$$

has a finite sum, when the very paradox is designed by dividing the tortoise's run into parts $\frac{1}{2}$, $\frac{1}{2} + \frac{1}{4}$, $\frac{1}{2} + \frac{1}{4} + \frac{1}{8}$, . . .

The point of the paradox could not be appreciated until in the first quarter of this century mathematics passed through a crisis comparable with that in Zeno's time. The critical question is this: does it make sense to speak of completing an infinite series of operations? E.g. does it make sense to talk of searching through all the numbers to see if there is an even number which is not a sum of two primes? Most mathematicians (then, I expect, as now) hold that it *does* make sense to talk of testing an infinity of cases. You can make the first test in $\frac{1}{2}$ second, the second in $\frac{1}{4}$ second, the third in $\frac{1}{8}$ second, and so on, and after 1 second you have made all the tests. Now I think this is what Zeno argued, that you can complete an infinity of steps, for Achilles does catch the tortoise and complete the infinite series of steps of halving the lead indefinitely often. The paradox is not that Achilles doesn't catch the tortoise, but that he does!

Now I think Zeno was wrong and that it does not make sense to talk of completing an infinite succession of operations. An infinite succession is without end, by definition. How then can we answer Zeno? Briefly, I think Zeno is confusing the infinite possibility of naming a point of subdivision of the route, with naming all the points. Thus if I draw a line from 0 to 1

---

0   1

I may be said to pass through infinitely many points on the way, in the sense that there is no limit to the number of fractions between 0 and 1 which I could name. But being able to name any one, and naming them all are two different things. Whatever I name, leaves some fraction unnamed. But even if we cannot name all the fractions, you may object, do we not *pass through* all the points in proceeding from 0 to 1? Suppose I count from 1 to 100 by 10's. When I count by 10's do I *pass through* all the other numbers, even though I do not count them? I might count by halves, but do I then pass through all the halves, under my breath so to speak, when I count by ones?

Zeno sought to defend mathematics against common sense. The present-day mathematician is on Zeno's side, but he makes no defence of mathematics. He simply denies that mathematics has to make common sense.

III

LANGUAGE AND EXPERIENCE

P ROBLEMS concerning the nature of signs and the relation of language to reality find expression in such questions as: 'Is language no more than a system of signs? Has language a *content*, or does it float above reality like a bubble above the earth? Can language point to something outside itself, has it roots in some actuality or are the truths of language independent of all experience? If language is a medium of communication (between human beings) then what is it that is communicated, and how is this communication effected?' As a preliminary to an examination of these questions, let us consider what we ordinarily say—and do—when we seek to decide whether a proposition is true or false. Suppose the proposition is 'The entrance to the University is in University Road'. To see whether or not this is true we might first seek the address of the University in a Street Directory, then we might ask various people the question 'Where is the University?' receiving the answer 'The University is in University Road', and lastly we might walk along University Road looking at each building in turn until we reach one bearing the name-plate 'University of Leicester'.

That the first two criteria are of the same character may be seen by supposing that the directory we consult is a machine which reproduces (from a record) the address of any institution when the name of the institution is spelt out on a dial on the machine. (For the present we ignore the sort of doubt which might be expressed by saying 'the man who is asked the question does not just *answer* (like the machine) but must think first before he can answer'.) When however we turn our attention to the third criterion we are inclined to think, not only that it is fundamentally different from the other criteria, but that it is conclusive in a way the others could not be; the reference book might contain a misprint, our informants might be mistaken or even wilfully deceive us, but 'we cannot doubt the evidence of our own eyes'. To see what this last criterion has in common with the preceding, we shall describe yet another form a reference book might take. The book might contain photographs of streets so that to find whether the proposition 'the University is in University Road' is true we look at the photograph of University Road to see if it contains a picture of the University. Thus we might contrast the criteria by saying that in the one we look at the object itself and in the other at a photograph of the object. Since we can use the reference book which contains the sentence 'The University is in University Road' and the reference book which contains the picture of the University in University Road' in exactly the same way, the sentence 'The University is in University Road' and the picture of the University in University Road, must stand to one another in the relation of syntactically equivalent sentences, like a French and English sentence with the same meaning. To translate from one word language to another we place side by side the words which may be written one for the other; to translate from a word- to a picture-language (like the directory containing photographs) we place side by side words and pictures. Thus the correspondence between a word and a picture language is established in the same sort of way as the correspondence between two word languages is established. But if we can find a correspondence between the sentence 'The University is in University Road' and a picture of the University in University Road, how can we doubt the existence of a correspondence between the sentence and that of which the photograph is a *photograph*, namely, the University buildings standing in University Road? Can we not translate from a picture language to a 'real object' language? If we want a man to build us a house can we give him only a picture of what we want (like the word 'house' or a photograph of a house) and can we not just point to a house? When a pure formalist maintains that there is no such process as deriving the truth of a proposition from some non-verbal occurrence, in what sense is he using the term 'verbal'? He accepts as a criterion of the truth of a sentence that the sentence forms one of a certain list of sentences. He affirms the possibility of deciding whether or not two rows of signs form the same *sentence*; for when he speaks of the consistency of two sentences, it is the shape of the signs which compose the sentences to which he refers, and he speaks of the possibility of changing from one notation to another, translating from one language to another. But if one admits the possibility of translating from one word language to another then one must admit the possibility of translating from a word- to a picture-language, and must therefore accept the *non-verbal* criterion of the truth of a sentence that a sentence is true if it is a translation of a picture ('sentence') that forms one of a certain collection of pictures. And if one accepts this criterion why should one not accept also the criterion of translation from an 'object language'?

Of course, the formalist may say that he meant no less than this himself, that by 'verbal' he meant anything that could be 'used as a word', but if this is so his statement loses its entire point; for if 'verbal' no longer serves to distinguish words from other signs, then 'non-verbal' has no meaning left to it, and if 'word' is being used in this new sense then to say that the world is the world of sentences is only to say that the world is the 'familiar' world of facts. We cannot deny the formalist the right to call the University standing in University Road a *sentence*; for some purposes this is a valuable form of expression, and harmless so long as we do not forget that it is a metaphor and not, as the formalists seem to imply, the expression of a new discovery about the world, the discovery that the world is a world of (what we used to call) sentences.

Talking to someone, sending someone a letter, communicating with someone, have the character of drawing someone's attention to something, holding something in front of someone, *pointing to something*. Pointing to a nut on a table and then pointing to the table may be called *saying* the sentence 'a nut is on the table' in the 'real-object' language, a sentence of which 'a nut is on the table' is the translation in the English word language. Saying what one sees, hears, feels, describing an experiment, recording an observation, are all translation processes. *Perceiving* a relation, *observing* a difference, *recognising* a likeness, are akin to *naming* a relation, a difference, a likeness. Pointing to a pencil and saying 'pencil' is one of the ways in which we translate from an object- to a word-language. Were it necessary, as the formalist maintains, not just to point but to say some such sentence as 'That which you see is called a pencil', learning a word language would be impossible; when a child is

C

taught to say 'sugar' each time it is shown a lump of sugar, it does not first have to understand the phrase 'That which you see is called . . .', we just attract the child's attention to the lump of sugar and say 'sugar', perhaps once, perhaps many times, and eventually the child says 'sugar' when we show it a lump of sugar. We teach the child to use the word 'sugar' as a token; if it wants a lump of sugar it must first give us the word 'sugar' in payment. What we teach is an exchange of *things*. Failure to understand this is one of the sources of the formalist's confusion; he feels that a definition must be a definition in words and accordingly he interprets the ostensive definition as defining the equivalence of the object-word and the ostensive definition sentence 'That which you see is called . . .'.

Bound up with the problem of the ostensive definition we meet one of the oldest of the problems of philosophy, the problem concerning the universal word. How is it possible, one might ask, for a child to learn that the word 'sugar' means, is a token for, any lump of sugar and not just some one particular lump? Must the child first perceive what various lumps of sugar have in common (whatever that may be) before it can learn to give the word 'sugar' as a token for any lump of sugar? Certainly, we could make a slot machine that would take only one particular coin, and reject all others, but we can also make slot machines that will take any penny piece, rejecting only coins of other values, or coins that differ in some other way from penny pieces. A child does not perceive what various lumps of sugar have in common, but fails to perceive such *differences* as there may be. Overlooking some differences in objects, but not overlooking others, is the fundamental operation in language. We regard a child's ability to learn languages quickly as a mark of intelligence, yet a too subtle and discerning child might never learn to speak his mother-tongue.

Let us examine more closely the three criteria we described above to decide the truth or falsehood of the sentence 'The University is in University Road'. We have already observed that the third criterion seems to be necessarily decisive, whereas the first and second are liable to error. Yet could we not conceive of the possibility of error also in the third criterion? If I walk along University Road and perceive a building bearing the name-plate 'The University of Leicester', may it not be that I am deluded and suffering from an hallucination? Is there in fact any criterion which is quite conclusive? Can we not doubt the validity of any criterion whatever? But if there is *no* criterion or combination of criteria that we are prepared to accept and call decisive, then the sentence 'The University is in University Road' is isolated from the language system and deprived of its function, like a currency that has lost its purchasing power. One might say that in choosing the criterion, the conventions according to which the sentence is true, or false, one is choosing the language in which the given complex of signs operates as a sentence. Accordingly, if we say that the reference book criterion for the truth of the sentence 'The University is in University Road' may be doubted, and, furthermore, that any criterion we conceive of may be doubted, then it is only in relation to some other criterion or group of criteria that a particular criterion may be said to be doubtful. If there were but a single criterion for the truth of a sentence, then it would make no sense to say that *this* criterion is doubtful; though in fact we do not ordinarily accept one criterion, we might do so. Language itself provides for the possibility of doubt as is shown by such words as 'mistake', 'falsehood' and 'hallucination', and the corresponding truth criterion of the assent of the majority. *They are said to be deluded who do not see what the majority see, whose world is not the world of the majority of men.*

It may seem to us, for we have grown accustomed to believing so, that only a madman could place any criterion above the criterion of experience, yet a few hundred years ago the criterion of the reference book (particularly the works of Aristotle and the Bible) was accepted in preference to the criterion of experience. That is not to say that men were then blind, ignorant or foolish (except according to our present criteria of knowledge). If what Aristotle said is *the* criterion of truth and if Aristotle said that a large object falls more swiftly than a small one, then Galileo was deluded, and however many times we drop objects of different weights, in a vacuum, and observe that they fall with the same speed, we are 'tricked by our senses' (and we might account for this trick in many different ways just as subtle as the theory of relativity). It is not a fact which is in dispute but the choice of a mode of expression. We do not dispute the fact that different objects are seen to fall with the same speed, the question is whether we shall use a language which says that the sentence 'Different objects fall with the same speed' is *true* because we perceive that different objects fall with the same speed, or whether we shall say that it is *false* and the perception a delusion.

The criterion which we called the criterion of experience, the criterion according to which the sentence 'The University is in University Road' is true if we see a building bearing the name-plate 'The University of Leicester' as we walk along University Road, might be formulated as a rule of logic permitting the derivation of the sentence '*p*' from the sentence 'I see *p*'. This formulation, however, raises the problem of the nature of such expressions as 'I see a red patch', 'I hear a ringing noise', 'I imagine a red patch', etc. We feel that these expressions have the certainty of necessary truths, yet they are neither linguistic conventions nor demonstrable sentences. The expression 'I see a red patch' has the conventional sentence form but does not play a sentence role in language; it makes no sense to ask 'How do I know I see a red patch?' for there is nothing which we should call the process of finding out that I see a red patch. Saying 'I see a red patch' is analogous, not to saying 'I have a red shirt' but to *painting a red patch*; in other words, saying 'I see a red patch', is like saying 'red patch'. The distinction which we express by 'I see a red patch', 'I imagine a red patch' is not a distinction between two activities but is rather the difference between painting a vivid red patch and painting a faint one.

Seeing a red patch and saying 'red', as opposed to 'defining' the word 'red' by pointing to a red patch, is a cause and effect phenomenon. Imagine what mechanism we may (association, the action of light on the eye, etc.), we cannot bridge the gulf between seeing the red patch and saying 'red'. It may help to make this clearer if we replace saying 'red' by painting a red patch. If a man looks at a red patch and then paints a red patch, is there a *logical* connection between what he saw and what he painted? Does it make sense to ask how he knows that the patch he is looking at and the patch he painted have the same colour? It is not whether he *may* be in doubt that puzzles us, but rather how he can fail to be in doubt. Suppose that on a shelf stand a number of bottles, each bottle bearing a label of a particular colour. I draw from a box a coloured token, place the token against each label in turn, reject bottle after bottle and then take down from the shelf the bottle which bears a label of the same colour as the token. How do I know that just this label has the same colour as the token? It cannot be necessary that I know what the colours are (called) since I may be unable to speak a word language, and if we say that I must *perceive* that the two colours are the same then in what does this perception consist but in taking down the bottle which I did take down? It might be objected that I must have seen *something*, else why did I choose just that bottle and no other, but what criterion have we for deciding this? I may say that I acted mechanically, that when I placed my token against that particular label I just reached for the bottle, and this may well be what happened; but that is not to say that I acted mechanically as opposed to consciously, for it is what we ordinarily call a conscious action that I am now tempted to call mechanical. We could in fact easily construct a machine which selected a bottle bearing a label of the same colour as a token placed in the machine; what puzzles us about the analogy with the machine is that we feel that when *we* choose a bottle *we* are guided by the *sensation of seeing the colours match*, whereas a *machine* cannot have sensations. Yet to say we are guided by our sensations is only to offer a hypothetical mechanism to account for our actions; for whatever sensation we experienced how could this sensation bridge the gulf between *seeing* the label and the token and *reaching* for the bottle?

Just as saying 'I see *x*', where '*x*' is an object word, is akin to saying '*x*', so saying 'I see *p*' 'I imagine *p*', 'I believe *p*', etc., where '*p*' is a sentence, is akin to saying '*p*'. Accordingly, the experience-criterion for the truth of a sentence may be expressed by saying that a sentence '*p*' is derivable from the sentence '*A* says *p*'. Of course, the sentence '*A* says *p*' may itself be derived from other sentences of the form '*B* says that *A* says *p*' and so on, and the choice of the initial sentence in the derivation process is quite arbitrary. Remember the legal convention that '*p*' is derivable from '*A* says *p* and *B* says *p* and *C* says *p*', but not from

'*A* says *p*'. Propositions for which the (accepted) criterion of truth is derivability from sentences of the form '*A* says *p*' may be called experimental propositions.

The reference book criterion for the truth of a proposition may be used in two essentially different ways. We might, for example, say that a sentence '*d*' is true if it is one of the sentences in Euclid's geometry or that '*d*' is true if it is one of the sentences in the first book on a certain shelf. Suppose that *a, b, c, d* are the sentences in Euclid's geometry so that the expression 'the sentences in Euclid's geometry' is synonymous with the class of sentences '*a, b, c, d*'; then '*d* is one of the sentences in Euclid's geometry' is derivable from '*d* is a member of the class *a, b, c, d*' which is derivable from the linguistic convention '*a, b, c, d* is the class whose members are *a* and *b* and *c* and *d*', so that '*d*' is true. In this case '*d*' does *not* express an experiential proposition; for the sentence '*a, b, c, d* are the sentences in Euclid's geometry' says 'the sentences *a, b, c, d* are *called* Euclid's geometry' and accordingly the sentence expresses a linguistic convention about the use of the expression 'Euclid's geometry' and is not a derivative of such a sentence as '*X* says that *a, b, c, d* are the sentences of Euclid's geometry'. But if we say that *a, b, c, d* are the sentences in the first book on a certain shelf, this is an experiential proposition and the sentence '*d*' which is derived from it expresses an experiential proposition.

The primary difficulty connected with the use of the criterion of experience may perhaps best be expressed by asking 'How do I know that you see what I see? Might not two signs seem to you to have the same form and different forms to me?' Is this a question about experience, or about reality, or about language? If we maintain that it is *impossible* to know whether you see the same thing that I see and if we refuse to accept any criterion according to which we should say that we see the same things, then the impossibility of which we speak is a *logical* not a *physical* impossibility. We say that no man can lift ten tons, and accept the test of men trying to lift the weight, and failing, and admit that of course a man may sometime in the future lift the weight. We do not say that any creature which lifts the weight shall not be called a man. Inability to lift the weight is not a defining characteristic of 'men'. But when we say that under no circumstances is it possible for *me* to know what *you* see, then it is with the use of the words *me* and *you* that we are concerned and not with the nature of experience. As Wittgenstein has observed, it is not *what* is seen that is in doubt but the choice of a language —whether we shall use the same word for what *you* see as for what I see, i.e. whether we shall admit both the sentences 'I see an *X*' and 'you see an *X*', or whether we shall allow the use of the object word *X* only in the sentence 'I see an *X*' and use some other word, *Y* say, in the sentence 'you see a *Y*' to express what is now expressed by 'you see an *X*'. But once we make this change we perceive that it is redundant; for the difference between the sentences 'I see an *X*', 'you see an *X*', is already clearly shown by the opposition of 'I' to 'you'.

Another form which the difficulties associated with this problem take may be expressed by asking 'How could I ever have learned the meaning of the sentence "I see a chair", for how could anyone else know *what* my sensations are when I say "I see a chair", and not knowing these sensations how could they have taught me to call just *these* sensations "I see a chair" ' (or, 'how could I ever learn the meaning of "toothache", for how could anyone else know when I have the experience *they* call "toothache"?'). Don't *I* mean something different when I say 'I see a chair' from what you mean when you say this sentence? I know what private experience I call seeing a chair, but I don't know what experience you call seeing a chair, nor even that you have any experience at all. Yet are the sensations I experience when I say 'I see a chair' the *meaning* of this sentence? If I behaved exactly as I now behave, brought you a chair when you asked for it, walked across the room without stumbling into the furniture, sat on a chair when I was tired, took my place in a row when you pointed to it, *and yet had none of the sensations which at present accompany my saying 'I see a chair'*, would you not still say that I understood the sentence? You might think that I could not behave as I do unless I have the sensations and experiences which I now have, but this is only a hypothesis. Imagine, for example, that the sensations I experience when I look at a red object and the sensations I experience when I look at a green object are interchanged, but I retain my present use of the words 'red' and 'green'; that is to say, I continue to use 'red' and 'green' in the way other people use these words, I stop my car when I see a red light, a red light still brings the word danger to my mind, and so on, even though the colour sensation I experience now is that which I experienced before on seeing a green light. I might notice the change myself in the sense in which I might notice that yesterday drinking cold water gave me a toothache, whereas to-day it does not, but the change would not be perceived in any other way. It could not have been my special *incommunicable* experience that I was taught to call red or I should now be obliged to change my language with the change in that experience.

Can a man show *all* the 'outward' signs of unhappiness and yet be happy? If he weeps and moans and presses his hand against his heart, rejects his food and speaks in a piteous voice of his grief—can this man, nevertheless, really be happy? Is it not possible, I might ask, that only my unhappiness is genuine and that others only simulate unhappiness; how can I know that another really feels as I feel when I say I am unhappy? But if I choose to say this and decide that the words 'I am genuinely unhappy' only make sense in my mouth, and if I now say 'he simulates unhappiness', where before I said 'he is genuinely unhappy', then to what is 'he is simulating' opposed? Consider the antithesis I formerly expressed by 'he is genuinely unhappy', 'he is simulating unhappiness'. My friend *A*, on receiving a letter telling him that his father has died, shows all the familiar signs of grief and mourning and another friend *B*, on receipt of the same news, gives similar evidence of great grief; I know, however, that *A* has always spoken affectionately of his father, imitated him in many ways, showed contentment in his company and expressed great concern over his illness, whereas *B* lost no opportunity to avoid his father's society, spoke disparagingly of him and impatiently awaited the inheritance his father's death would bring him. I should say that *B* only simulated unhappiness but, I should say this from what I know of the 'context' of his grief, not because I know of some private sensation which *A* experienced but *B* did not.

If we admit the possibility of doubting any truth-criterion, that is to say if we maintain that logic leaves us free to choose any language we please, we seem to lose the connection between language and reality which we thought the concept of truth established, or rather if we maintain that connection, then reality loses its uniqueness and is set free to revolve alongside the turning wheel of language. The correspondence of language and reality takes on again the character of an illusion, for the correspondence subsists only so long as we build the world in the image of our language. 'Was that', runs a Chinese aphorism, 'Lao-tse dreaming he was a butterfly or this a butterfly dreaming he is Lao-tse'. I see a piece of wood before me. I put out my hand and touch it, feel the contact in my finger tips, see my fingers touching the wood. I take a saw and saw through the piece of wood, watch the dust falling, smell the fragrance of pine, observe the grain, indent the surface with my thumb-nail, place the sawn-off piece in a basin of water and watch it float. Do these things prove that wood is a *real* substance? Might I not just be watching a private cinema show, or dreaming? And if others talk to me and tell me they too see the wood and smell the pine and hear the rasp of the saw, might not all this too be part of my dream? Can I ever be sure that I have found *reality*? Yet why do I use the word *real*? I know perfectly well how I distinguish between a real piece of wood and an artificial piece of wood, how I distinguish between sawing a piece of wood and dreaming that I am sawing a piece of wood. I use the familiar words, but I want to give them a new and special sense, and it is essential that this new sense be private to me alone, that I cannot communicate this sense, for if I could communicate it I should have to draw a distinction between *real* and something else, and this I don't want to do.

A real object, one might say, has a length. What is the length of this bar of iron? I place my standard (millimetre) measuring rod alongside the bar of iron and find that I can set it off on the bar between 120 and 121 times. Accordingly, I say that the length of the bar is between 120 and 121 millimetres. You object and say that I omitted to take into account the temperature. I repeat the operations (obtaining the same result) and write alongside my result the temperature 30° Centigrade recorded by my thermometer. Next day, when the thermometer records 35° Centigrade, I find I can set off my measuring rod between 121 and 122 times along the bar. What is the *real* length of the bar of iron? We should answer, without hesitation, that there is no sense to the question; for the length of the rod depends upon the temperature, unless we choose some particular temperature as a standard and call the real length of the iron bar the length associated with this standard temperature. If we observed, however, that whatever the variation in temperature (between some assigned limits perhaps) the length of the bar was always between 120 and 122 millimetres, we might call 'between 120 and 122' the *real* length of the bar and this would be in effect a change of unit, for if our standard measuring rod was 2 millimetres long we should now record the length as between 60 and 61 units.

Suppose, however, we said 'keeping the temperature constant, surely the rod must have an *exact* length to which the measurement 'between 120 and 121 millimetres' is only an approximation'. If by this we mean only that with no matter what unit we carried out the measurement, we could imagine it carried out with a smaller unit, then it is true; but if we mean that we shall call 'the length of the bar' not the result obtained by any measurement but only the 'limit' towards which the successive measurements converge, then we have deprived the word 'length' of its use; for however many measurements we carried out there would remain the unlimited possibility of carrying out further measurements and the length of the bar is now *unattainable by definition*.

If we accept some other criterion than measurement for the determination of the length of the bar, for instance some calculation based upon the velocities of the end points of the bar, then we are giving the word 'length' a new use and must be prepared for the possibility of a different answer to the question 'What is the length of the bar of iron'. Thus when the relativitist says 'That which you thought to be a bar of fixed length is really shorter when it lies in some positions than others, only it is impossible for you to detect this difference as your measuring rod also changes its length and your physical organs change in such a way that you are not aware either of this change or of the changes in the lengths of objects', he is using the words 'length' and 'change of length' in two different ways. Has the relativist shown that the ordinary man's use of 'length' is wrong? Is the relativitist's answer to the question 'what is the length of the iron bar?' the *true* answer and the answer which the measuring rod gives *false*? Was it just a foolish prejudice, a habit of thought, to think that we can measure lengths with a measuring rod, a prejudice from which the relativitist sets us free? The relativitist mistakenly expresses a new convention about the use of the word 'length' in the form of a discovery about the nature of the world. It is not a new fact that the relativitist records but a change of language. If we mean by 'length', as we ordinarily mean, that which is determined by a measuring rod, then it makes no sense to talk of a change in the length of the measuring rod, for the measuring rod (the standard of length) has no length. By changing the meaning of the word 'length' we appear to make the strange discovery that that which we thought to be the instrument of measurement in some way has now been shown to measure itself.

The effect of radium on living tissues is a new fact, the Copernican astronomy a new language. In the history of the human race the discovery of a new language may be of greater importance than any discovery of the experimental sciences, but for the philosopher a new language is *only* a new language, valuable for purposes of comparison, but having no greater claim on his attention than any other language. The Copernican or the relativitist use of the word 'motion' is just *one* of the uses of the word. There is a common belief that the Copernican use of the term 'motion' was forced upon us by the overwhelming evidence of facts, that in some sense we were driven to 'admit' the motion of the earth which ignorance had hidden from us; whereas what Copernicus discovered was no new fact of the Universe but that the paths of the heavenly bodies were *more simply described* if by motion we meant motion round the sun and not motion round the earth.

Parallel to the question 'what is the nature of reality?' we might ask 'what is the nature of experimental science?' or 'what validity has the scientific method?' We photograph the motion of a planet across the sky during some interval of time; the photograph presents the motion of the planet as a strip of light across a darkened background. We wish to prolong this strip of light, but of the unlimited possibilities of prolongation, there is no one particular prolongation that is logically necessary. If two photographs are taken, one for six months and the other for this and a further six months, can we, without looking at the second photograph fill in on the first 'unfinished' photograph the path that the second photograph is recording? That is the problem of the experimental sciences. Theoretical physics singles out some particular one of the unlimited possibilities of prolongation as *the* correct prolongation, the 'reasons' given for this choice constituting the so called theory of the science, and this prolongation is compared with the second photograph. If the prolongation is a good fit, other experiments of a similar character (choosing a prolongation according to the theory) are carried out, and if no marked discrepancy between the 'theoretic' prolongation and the second photograph is observed, the theory is said to be well founded and provisionally adopted.

If we look more closely at the 'reasons' given for a particular guess (theoretic prolongation) we find a mixture of observational and linguistic sentences. Some reasons, like 'the velocity of light is independent of the velocity of the source of light' are linguistic statements, expressed in the form of observational statements. 'The velocity of light is independent of the velocity of the source of light' tells us how the expression 'the velocity of light' is going to be used. Such a statement lays down the way we have decided to talk about our experiments, serves to choose between conflicting evidence. Just as a principle of law like 'a man is innocent until he is proved guilty' tells us, not something about the character of human beings (that a man cannot have done wrong unless he is proved to have done so), but something of the way we are going to use the words 'innocent' and 'guilty' and serves to guide our *treatment* of untried prisoners. Moreover, even as we are prepared to change a principle of law—i.e. to adopt a new legal language—we are prepared to change the reasons we give for our guesses (apart from changing the guesses themselves).

One might say that theoretical physics is a link between language and experience, the so-called theory forming a dictionary for translating observational sentences into mathematical equations and vice-versa. Consider the statement 'Forces are added by the parallelogram law'. Is this an observational statement? Certainly we can use this rule to build bridges and certainly bridges built according to it have 'stood the test of time'; but might not a bridge built to-morrow according to the rule, fall down when tested? Contrast the statement with 'vectors are added by the parallelogram law' which is the mathematical *definition of addition* for vectors. We can imagine a time when men perceived how forces are added, and stated that forces add by the parallelogram law just as one might observe that twenty people pass one's window every day. At such a time the parallelogram law is a statement of fact, not a generalisation from particular instances, not an induction, but a statement of fact. At some subsequent time the statement of the parallelogram law ceased to be used as a statement of fact and became a linguistic convention. Though certain experiences in the future may lead us to abandon the use of the word 'force' given by this convention, we might retain the convention in spite of any experience whatever. In the same way the propositions of arithmetic 'evolved' from observational truths to linguistic conventions. It might happen in the future that when we placed books on a shelf or apples in a bag or men in a room and counted them one by one we never found a total beyond five; that is to say, after counting the books and finding there are five on the shelf, and then placing further books on the shelf and counting again, we still obtain the answer five, and similarly after counting and recounting apples and men and so on. In such a world we might lose our interest in our common arithmetic (which of course is not invalidated by such experiences any more than it is validated by our present experiences) and adopt instead an arithmetic in which the sum of five and one is five. On the other hand, we might still retain our common arthimetic and say that there are really six or seven, etc., books on the shelf but some have coalesced so that we seem to have only five. Compare this with the position in which present day physics finds itself. Shall we say there are material particles which behave like waves, or waves which behave like material particles?

The microscope shows a drop of water as a universe of active and brightly coloured creatures, of forms unfamiliar to the naked eye. Which is the real drop of water, that which the unaided eye shows us or this myriad of tiny creatures? Does the microscope enlarge our knowledge of reality or does it destroy the possibility of belief in any world of the senses? If we answer that there are two worlds, the world of our (ordinary) vision and the microscope world, which of these is the real world? And perhaps the creatures of the microscope would likewise under some more powerful microscope reveal themselves just as organisations of other (smaller) creatures. Related to these questions is the metaphysical problem 'are not human beings just parts of some greater organism of which we can have no conception?' and the allied problems such as 'have human beings an independent existence or do they just serve some higher organisms, call them nations or civilisations, even as we perceive that the creatures which dwell in our blood live but to serve us?' Another form in which we can express these questions is 'are microbes real?' or 'is there really a microbe in this drop of water?' It is important to remember that if we say microbes can only be seen through a microscope, they are too small to be seen with the naked eye, we are making a hypothesis, not stating a logical necessity of vision. It is a hypothesis that *what* we see (when we look through a microscope) depends upon the microscope. We can imagine a world in which on some days our visual experiences are what they are now when 'we look around with the unaided eye', and on other days are what we are now familiar with on looking through a microscope. What we have before our mind's eye when we ask if there are really microbes in a drop of water is an image of a clear drop of water beside an image of the same drop speckled with microbes; but looking at a drop of water and then looking at the same drop through a microscope is more like seeing a clear drop of water transform into, not a speckled drop, but a vessel of water filled with tiny creatures. That this vessel of water and the drop of water are the same thing is also only an hypothesis; that is to say the form of expression 'this vessel of water you see is really the drop of water seen through a microscope' expresses a purely arbitrary convention, one which we could abandon without denying any fact. Instead of saying that the microscope shows us the microbes (hidden) in the water we could say that the microscope *transforms* the drop of water into an expanse of water filled with swimming creatures; of course, this change of expression (remember it is not the facts we change) will entail other changes of expression. Instead of saying that a man has typhoid fever because the microscope reveals the presence of the typhoid bacillus in his blood, we should say that the microscope transformation of his blood contains the typhoid bacillus. And to the question 'surely there must be something already in the blood (before the microscope transformed it), else why is the man ill?' the answer is that to say the man is ill *because* of a microbe in his blood is only to make a hypothesis—a linguistic convention—which we can retain or abandon as we please. If the microscope transformation of my blood is found to contain the typhoid bacillus, I should unhesitatingly accept the treatment which I had observed in previous cases to be followed by a return to health, not because I had found out that 'such and such a process *cures* typhoid' is more than a linguistic convention, but because that is how human beings behave.

It is sometimes said that just as there are microbes too small to be seen by the unaided eye, so too there are electrons too small to be seen even by a microscope. This is rather like saying that something is seen which is not seen. The sentence 'I see an electron' has not that relation to 'I see a microbe' which this latter has to the sentence 'I see a drop of water'. If we ask a physicist to show us an electron he shows us pictures of white (or coloured) lines on a black background and calls these pictures electron tracks; the physicist will say that he cannot show us single electrons, only streams of electrons, and will in fact show us not streams of anything, but just streams. It may be true that nowhere in the world can we find a single locust, that locusts are found only in swarms, but a swarm of locusts is specifically a swarm of *locusts*. The physicist tells us that he can distinguish one electron from another and in support of this shows us pictures which he calls 'streams of α-particles', streams of β-particles' and so on, but it is not the contents of the streams by which these pictures differ. Again we are tempted to ask the familiar questions 'are electrons the framework of reality, is the world of electrons the real world behind the illusory world of the senses?' 'Is the real chair that which we see and handle, buy and sell, cover with tapestry or chop up for firewood, or is the real chair an organisation of electrons?' Is it a mark of ignorance to believe there are no electrons; does not the electric light prove the existence of electrons? Are there facts which we cannot describe without the use of the word electron (or a syntactically equivalent word)? If we say that the supposition that there are electrons explains the phenomenal world and enables us to predict the future so accurately, that it is no longer possible to doubt the truth of the supposition, do we mean that because of this success (in foretelling the results of experiment, etc.) we have more confidence than before that we shall one day 'isolate an electron'? or do we only mean that we lose any temptation we may have had to abandon electron-language? After reading in the paper one morning that a certain man has taken his life and that beside his body a letter was found saying he was going to shoot himself because he had lost all his money, I might write a play in which a man immigrates to this country, works hard, makes a fortune, loses it and shoots himself, and then find that I have described correctly in every detail the life of the man I read about in the paper. Suppose that I repeated the experiment many times, that each time I read of a suicide I write a play and then find I have correctly told the life story of the dead man. If I now say that I believe I have second-sight, will my success have proved the truth of my supposition? Of course, if we agree that 'I have second sight' is just a form of expression for describing what I did, then to say I have second sight is just to say that I did what I did; but could we not dispute as to whether there is such a thing as second sight? And if someone says that *nothing* will convince him of the reality of second sight, is he denying the possibility of a certain experience, or just refusing to use a certain form of expression? On the one hand I certainly did not *know* the life story of the man who took his life, yet I wrote that story correctly in every detail. The facts are not in dispute. I did not know the story but what I wrote turned out to be written just as if I had known the story. (The scientist does not know when there will be an eclipse, but what he says turns out to be true just as if he had known). Shall we give a description of these facts which *stresses* the similarity of my writing to actual knowledge or shall we give a description which minimises this similarity; that is the choice we make when we accept or reject the expression 'second-sight'. If we grant the man who says he believes in survival after death any experience that he may desire, even a body walking the earth alike in all respects to that of one who has died, having all the memories of the dead man, speaking with the same voice, behaving in all respects like the dead man formerly behaved, must we say that the survivalist's belief has been proved true? Is there a fact in dispute if we argue whether we shall say the dead man has returned to earth or whether we shall say that the man who died has not returned but another exactly like him now walks in his place?

The word 'reality' has whatever significance we choose to give it. To say that there is no correspondence of language with reality is to make the decision not to use the word 'reality' and to express this decision with the air of making a new discovery about the nature of the world. And if we ask whether there is not some *Reality* behind the phenomenal world, a reality of which the world of our senses is but a shadow, we must answer that if we choose to change our language, and talk, not of tables and chairs, but of shadows-of-tables and shadows-of-chairs and call what we now term 'shadow' instead 'shadow of a shadow', then we perceive that 'shadow-of-a-chair' is now a complex term, that for the word 'chair' itself we have given no use and therefore 'shadow-of-a-chair' is but a redundant form of expression for 'chair'.

D

IV

THE MEANING OF COUNTING

BEFORE men learnt to *count*, the shepherd guarded against the loss of a sheep from his flock by making a *tally* of the flock. As each sheep passed through the gate in the morning, on the way to the grazing land, the shepherd cut a notch in his stick. At nightfall, when the flock returned to the fold, he checked his sheep against the tally by running his finger along the notches, moving from notch to notch as each sheep passed. The tally stick served its purpose just as well as counting the flock serves the shepherd today. The tally could be passed from hand to hand almost as easily as numbers are now passed by word of mouth (or written down); it constituted a reasonably permanent flock record, and so long as flocks remained quite small there was no necessity to mother the invention of a new system of recording.

A very simple modification of the primitive tally stick extended its scope to a remarkable degree. Two sticks, in place of one, served not just to double the size of the flock of which a tally could conveniently be made, but to increase it perhaps twentyfold. One of the sticks, serving as a standard, carries, say, twenty notches, and the second, separated into two parts by some distinguishing mark, is initially uncut. This modified tally-stick is used in the following way. As each sheep of the flock passes through the gate the shepherd moves his finger up one notch on the standard stick, and when he comes to the top of the standard, he cuts a notch on the lower part of the second stick, and then starts again at the first notch on the standard. Each time his finger reaches the last notch on the standard he cuts a notch in the second stick. When the tally is complete the standard is laid alongside the upper, uncut, part of the second stick and the notches on the standard, up to the last covered in making the tally, are matched against fresh notches cut in the second stick. The second stick now serves just as well as the primitive tally as a guard against loss from the flock, for the process of construction is reversible with the aid of the standard stick; on the return of the flock the shepherd moves his finger from notch to notch on the lower half of the double stick each time he completes a tally on the standard stick, and when the last notch on the lower part of the stick is reached, the standard is discarded and the remaining sheep are matched against the notches in the upper part.

This tally-of-a-tally, as we may describe the two-stick tally, effects an immense economy both in effort and material. For instance the tally of a flock of four hundred and seven sheep will consist of only twenty-seven notches. The number of notches on the standard is, of course, quite arbitrary, and a standard of twenty has no special advantage; the ten fingers on our hands, however, provide a standard so readily accessible and convenient that a standard of ten was adopted almost universally.

The *abacus*, which is a familiar toy in the nursery to this day, is a development of the two-stick tally. The prototype of the abacus consisted of a few vertical wires open at the top to allow the free passage of beads off and on the wires. To simplify the description of operations with an abacus we shall suppose the abacus to be so set up before us that we can talk of a right-hand and left-hand end wire, and we shall think of the wires as ordered from right to left, so that by the term 'the next wire' we mean always the next to the left, and by 'the previous wire' we mean the next wire to the right.

In making a tally on the abacus the beads take the place of the notches, and the wires that of the sticks. A two-wire abacus is used in very much the same way as the two-stick tally, the addition of further wires serving to accentuate the economy in expression gained in replacing the primitive tally by the two-stick tally. For the moment we shall suppose that each wire is just long enough to carry ten beads. When the first wire is full a bead is placed on the second wire and the first is then emptied; the next bead is again placed on the first wire, and so on. When the second wire is full a bead is placed on the third wire, and the second is emptied, and the next bead is again placed on the first wire, and when the third is full, a bead is placed on the fourth wire and the third is emptied, and so on. Each wire in turn serves as a standard, the beads on a wire recording the number of times the previous wire has been filled. In reversing the process, we start at the right-hand end, and when the first wire has been emptied a bead is removed from the second wire and the first wire is filled again. When the second has been emptied a bead is removed from the third and the second wire is then filled again, and so on.

After the invention of the abacus the next stage in the development of Arithmetic was the discovery of a method of recording (on sand or paper) a tally made on an abacus. We are so familiar with the process of counting that it is exceedingly difficult for us to realise how deep a problem this recording presented before the invention of counting. The abacus itself takes us a long way towards a solution, for by distributing the beads amongst several wires, a *positional notation* is strongly suggested. In fact the problem is reduced to that of finding a method of recording the (relatively few) beads on a single wire, the complete tally being then expressed by the juxtaposition of the symbols standing for the beads on the single wires. A mere *copy* of the beads on a wire will not suffice for our purpose, for this simply replaces one abacus by another (or by a drawing of an abacus) and so constitutes no real development. The essential feature the recording must have is that the result be presented in a form which is *recognisable at a glance*. The problem was solved by arranging the beads in simple patterns. One method of arrangement (which was in part used by the Greeks) is shown in the next line

.   ..   :.   ::   ::.   :.   ::   ::   :::

The weakness of this arrangement, from the psychological point of view, is that the differences between the patterns, particularly the last five, are not sufficiently striking. Another factor which contributed to their falling out of use is their lack of 'cursiveness', which made a rapid recording, on sand or paper, impossible. But, historically, their introduction marks a decisive step forward. In conjunction with the abacus it rendered possible a most valuable abbreviation of a tally and facilitated the recording of large flocks in a simple and easily recognisable form. Here is an illustration of their use. In the first picture we

have a tally made on an abacus, and above it is the record of the
tally made in a positional notation.

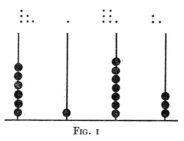

FIG. 1

Once the use of such number-patterns has been adopted, two
obvious lines of development are open. On the one hand, we
might give names to the individual patterns, the names then
taking the place of the patterns, or we might introduce (in the
course of time) some system of symbolic abbreviation of the
original patterns. A simultaneous development along both these
lines took place (though in part divorced from the positional
notation of the abacus). This development was based upon an
existing alphabet, with letters in an established order of
precedence, and is logically irrelevant. The successive patterns
were both called after and represented by the letters in their
'natural' order. For the sake of illustration we may take the
Roman alphabet as a model, though in fact only older alpha-
bets were actually used in this way. *Translation* from dot patterns
to alphabet signs is carried out by means of a table in which the
pattern and its letter are placed side by side. For convenience in
printing we place the entries in horizontal rather than vertical
columns though the latter would be more 'natural'.

a	*b*	*c*	*d*	*e*	*f*	*g*	*h*	*i*

Thus, for instance, *f a g c* stands for six thousand, one hundred
and seventy-three.

The transition from dot patterns to alphabet signs is logically
irrelevant, for, as we shall see later, the development of arithmetic does not presuppose the existence of an alphabet, and in
fact numbers themselves afford a more useful index of ordinal

relations than does an alphabet. None the less the transition is important for reasons which we must now consider.

The dot patterns are not simply an adjunct to an abacus, for once their use has been understood the abacus itself may be dispensed with. Notice first that the construction of the pattern is *progressive*, each pattern being incorporated in the succeeding one, as the following diagram shows.

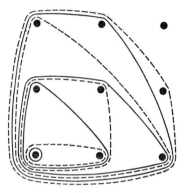

FIG. 2

Accordingly, in making a tally of a flock, we may place a pebble on the sand for each sheep that passes us, starting with one pebble, then two, and so on, each additional pebble being so placed as to transform the existing pattern into the next. Conversely, taking away one pebble at a time, the pattern may be unmade in such a way that after each removal of a pebble another dot pattern is left, until the last pebble is reached. Making a tally by dot patterns differs, however, in one small respect from using an abacus, or rather differs from the use of the particular abacus we described. We supposed that each wire of the abacus holds ten beads, whereas only nine can be represented by the foregoing dot patterns, but in fact it would be sufficient for a wire to hold only nine beads, since ten is represented by a single bead on the second wire. The nine-bead abacus is operated by placing the *next* bead on the second wire when the first wire is full and then emptying the first wire. Correspondingly when a flock is being checked against a tally made on a nine-bead abacus a sheep must be matched against the last bead on the second wire *before* the first wire is filled again. In the ten-bead abacus a bead on the second wire *represents* a full first wire, that is to say, may be replaced by a full first wire, but this is not the case with the nine-bead abacus, where a bead on the second wire is the successor, not the representative, of a full first wire. The ten-bead abacus is in fact the simpler to understand, but on account of the dual representation of powers of ten, is logically more complex than the nine-bead abacus. The dot patterns have the same multiplicity as the nine-bead abacus.

Just as the dot patterns may take the place of an abacus, so do the alphabet signs, at first the *names* of the dot patterns, subsequently acquire an independent usage. Instead of *constructing* dot patterns we operate directly with the alphabet signs, their natural order being supposed known and taken for granted. The succession

.   ..   ⦂   ⦂⦂   ⦂⦂   ⦂⦂.   ⦂⦂.   ⦂⦂⦂   ⦂⦂⦂

becomes, simply, *a*, *b*, *c*, *d*, *e*, *f*, *g*, *h*, *i*; as each sheep passes through the gate we say (or write) in turn the alphabet signs, the use of a positional notation extending a relatively small alphabet to an unlimited extent.

In considering positional notations we have not yet mentioned the problem presented by an empty wire. For instance, thirty-one is represented adequately by *ca*, but how shall we represent thirty? If we write just *c* there is no means of knowing whether this stands for three, thirty or even three hundred, since we have not indicated the empty first wire in any way. Looking back at the problem from our present standpoint it seems obvious that all we require is a sign to denote an empty wire—an office which is performed for us by the letter o—but historically and philosophically this problem was by no means trivial. Remember that the letters are being used to denote dot patterns; how then, it was asked, can a letter also represent the absence of a dot pattern? (How can nothing be a number?) Yet can we not picture a wire without beads, just as well as a wire on which beads are threaded, and may we not allow some trace of

the abacus wire to persist in our new notation? The problem of
the zero arises only in passing from an abacus to a symbolic
representation of an abacus, and is therefore only a pseudo-
problem, a problem in *notation*. The dot patterns, or the alphabet signs, supplemented by the zero, o, form a complete
representation of the abacus.

The transition from dot patterns to alphabet signs takes us
from the tally to something fully equivalent to counting. The
essential difference between making a tally (for instance, on a
single stick), and counting (by alphabet signs) is the difference
between longhand and shorthand (almost the difference
between travelling and arriving). The tally stick shows not only
the flock but also all the individual parts which make up the
flock. Alphabet counting records the flock as a whole, completely obscuring the track by which the total was obtained. The
tally stick is like a *copy* of a flock, and the alphabet-count like a
label on a box, a summary of the contents. Of course, in its
primary role of *naming* dot patterns the use of alphabet signs in
no way transcends the tally stick, no more than any translation
transcends the original; it is the use of alphabet signs, *per se*,
divorced from dot patterns that constitutes the invention of
counting. And this usage, let us repeat, is founded in a supposedly previously established natural order of alphabet signs.

Before turning to the modern system of counting which super-
seded the use of alphabet signs, there are still a few points of
interest to be noticed in connection with the abacus and the
positional notation. We have already observed that although
our two hands give the number ten an obvious 'natural' advantage over any other number as a base of a positional notation,
there is no logical reason for preferring one number to another.
A large base number, that is, a long-wire abacus, represents
large numbers on comparatively few wires, but to offset this
advantage renders the transition to dot patterns more difficult,
involving as it does the perception of relatively complicated
patterns (think of a system of dot patterns for numbers up to
twenty). On the other hand, a small base number like two (the
base number cannot be less, since an abacus which admits only
one bead on each wire is just a primitive tally) whilst reducing
the number of dot patterns to a minimum, requires as many as

seven wires for the expression of even such a small number as sixty-four. There is, furthermore, no logical necessity for the wires on an abacus to be of equal length, and though an abacus with wires of varying lengths may have no obvious practical advantage, its use is no whit more complicated than that of the more familiar kind, and, what is more important, its possibility helps to break some of our old-established prejudices regarding the nature of a positional notation. For instance, we might construct an abacus in which the first wire carried only one bead, the second only two, the third only three, and so on. Its usage will differ in no respect from the nine-wire abacus we have already described. The following diagram illustrates the representation of the numbers one to nine on this abacus.

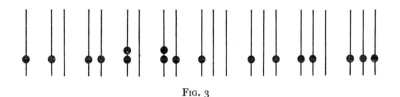

FIG. 3

A positional notation is by no means the only literal translation of an abacus tally. Apart from the use of various arithmetical operations, which we shall consider later, there are other methods of indicating the abacus wires than the *positions* of the dot patterns, or letters. For instance, we may use the dot patterns twice, both to denote the number of beads on a wire, and also to denote the wire itself. First, wire by wire, we make tallies of the wires in dot patterns (notice that we may treat the wires themselves as a primitive tally) and in this way each wire is associated with a dot pattern, which is in fact a tally of all the wires not on its left. Then we arrange the beads on each wire in a dot pattern placed below the pattern associated with that wire. Thus two thousand and sixty-three is represented by the complex

$$\begin{array}{cccc} \because & \because & \cdot \\ \cdot & \cdots & \cdot \\ \cdot\cdot & \because\cdot & \because \end{array}$$

and ten thousand and forty by

$$
\begin{array}{cc}
\vdots\; : & \;\cdot\cdot \\
\hline
\cdot & :\,:
\end{array}
$$

Observe that this system of representation has no need for a sign representing an empty wire, and therefore, occasionally, effects some economy in representation. If we replace the dot patterns by alphabet signs (and distinguish wires and beads by the use of capital and small letters) we obtain the complexes

$$
\begin{array}{ccc}
D\; B\; A & \qquad & E\; B \\
b\; f\; c & \qquad & a\; d
\end{array}
$$

which are practically equivalent to our present representation of numbers *in words*. For though we do not in fact employ letters, the *system* is almost the same; we call the wires, not A, B, C and so on, but units, tens, hundreds, and so on, and then specify the number of beads on each named wire, ignoring empty wires. The difference between the systems, expressed in modern ter-minology, is this: in the alphabet notation we number the wires one, two, three and so on, in turn, but in the current notation a wire is called after the number represented on an abacus by placing just a single bead on that wire. The first wire is called the unit, since a single bead on the first wire represents a unit, which is another word for 'one', the second wire is the tens wire since a single bead on the second wire represents the number ten, and so on. That a hundred is ten tens or a thousand ten hundreds is not a theorem (on multiplication), but just an expression of the rule that (on a ten-bead abacus) a bead on one wire may replace or be replaced by a full wire to the right. Interesting questions arise when the number of wires in an abacus exceeds the number of beads which fill a single wire, for the alphabet notation breaks down; it can, however, readily be extended. For instance, since the alphabet notation, as described above, enables us to record any number up to $10^9 - 1$, that is,

$$
\begin{array}{c}
I\,H\,G\,F\,E\,D\,C\,B\,A \\
i\;\; i\;\; i\;\; i\;\; i\;\; i\;\; i\;\; i\;\; i,
\end{array}
$$

by using complexes instead of pairs of letters the notation will cover an abacus of $10^9 - 1$ wires. For instance, $10^{12} + 1$ is represented by

$$\begin{pmatrix} B & A \\ a & c \end{pmatrix} A$$

$$a \qquad a.$$

Thus extended, the alphabet notation will represent any number up to

$$k' = 10^k - 1, \text{ where } k = 10^9 - 1.$$

In a similar way, by adding a fourth row, we can next extend the notation to represent any number up to $10^{k'} - 1$, and so on.

The present system of numeration derives not from dot patterns, but from stick patterns. If we make a tally of a collection (of not more than nine objects) by laying sticks in a row, the tally may be rearranged into one and only one of the following patterns:

These stick patterns are both more striking, and individual, than the dot patterns, and lend themselves very readily to a cursive reproduction on sand or paper. Compare these patterns with the arabic numerals as we know them today:

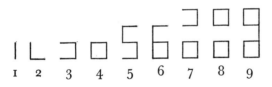

Though we have no historical record of these stick patterns the correspondence exhibited in these two rows is itself very strong evidence of a stick pattern ancestry for the present-day designs. Variations are of course possible. It may be that the numeral *three* was at one time reversed, and perhaps also the numeral *nine*. The seven and the nine have lost the lower loop. But none of these changes is more than one would expect to find taking place in the course of a transition from square to cursive patterns.

The stick patterns are not contained one inside the next, which was a remarkable feature of the dot patterns. But this disadvantage is more than offset by the ease with which the patterns can be identified and reproduced. Moreover, these patterns may have descended from an older set, which exhibited this property. For instance, the following:

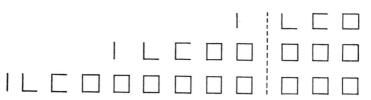

Here all the signs 'look to the right', and each is contained in the next. They are, however, clearly constructed on the base four, and it is hard to believe that there was a general use of this base antedating the introduction of the base ten.

From each stick pattern we obtain the next by the addition of a stick, and the perception of this is the foundation of counting. By naming the individual patterns (or replacing them by symbolic abbreviations like our numerals), we pass from tally making to counting, as we have already seen in regard to dot patterns. The process of transition may be imagined in a number of stages. First the tally is made by laying sticks in a row, one for each object of the collection of which the tally is being made. Then the complete tally (which we suppose for the moment to be of a collection of fewer than ten things) is arranged in a stick pattern, and the name of this pattern is recorded. This may be regarded as the frontier in the development of arithmetic across which we pass over to counting. So long as we delay the construction of a stick pattern to the *last* stage we have not crossed this frontier. Now suppose that, instead of laying the sticks in a row, we form a stick pattern each time we lay down a stick, the last pattern formed alone being retained. *This is counting*, as we recognise immediately we substitute spoken words for stick operations. Name each pattern as it is formed, and say 'and one' for each fresh stick we lay down (and so for each object of the collection) and we shall be accompanying the process of pattern construction by the refrain

one, and one is two, and one is three, and one is four, etc.,

where we have used the English names of numerals as stick pattern names.

Of the two, stick pattern construction and verbal refrain, one is superfluous, for each may stand alone. The one by one passage from each stick pattern to its sequent may be expressed in words alone, without carrying out the actual pattern construction; instead of laying the stick alongside the three-pattern to make the four, we just say 'three and one is four', and after each 'and one' we name the next pattern, without troubling to construct it. In the course of time, we omit the reiterated 'and one', and instead perhaps we touch an object, or just look at it. In this way counting becomes apparently a mere recitation of stick pattern names, one, two, three, four, and so on, which is what makes counting seem at first sight such a mysterious process. But, as we have seen, counting is *not* a recitation, but a process of *transforming a tally into stick patterns*, or, rather, a description of this process without actually carrying out the pattern construction. What in fact we are saying when we count is, let us repeat, 'one, and one is two, and one is three, and one is four', etc., though each 'and one' may be expressed otherwise than in words, for instance by moving along the objects we are counting, or even by glancing at them one by one.

With the help of a positional notation (or one of the other devices we have already described in terms of dot patterns or alphabet signs) the extension of counting beyond the range of the nine stick patterns is readily effected. We need only names for the abacus wires (that is for the successive positions occupied by the patterns in a positional notation) and these names are the familiar tens, hundreds, thousands, etc. Counting beyond nine proceeds thus: nine, and one is ten, and one is ten-one, and two is ten-two (familiarly contracted to eleven, twelve), etc.

The nomenclature for the successive positions occupied by digits in the positional notation is (apart from the initial positions) capable of systematic extension. The terms units, tens, hundreds, thousands, follow no rule, but are simply specific names for the first four positions. From thousands to millions a serial law operates, giving rise to the terms 'tens of thousands' and 'hundreds of thousands' (utilising the already prepared term 'hundreds' for 'ten tens'). From a million, which is a thousand thousands, we advance through ten-million, hundred-million, thousand-million, ten-thousand-million and hundred-thousand-million to a billion (a million-millions). To prolong the nomenclature we require a name for n n's where n is the last name introduced. Thus, just as we have a million for a thousand-thousand, a billion for a million millions, a trillion for a billion billions, then the indefinite prolongation we are seeking may be obtained by adding the ending 'illion' to the name of a number $n+1$ to denote n-illion n-illions. For instance, decillion for nonillion nonillions, or million-one-illion for millimillion milli-millions. (Since we are here concerned only with possibilities, not with actual usage, we need not attempt to find more melodious substitutes for these cacaphonous hybrids.) Alternative, and simpler, systematic notations are readily devised if we ignore the indications of the current nomenclature. For instance, we may denote the successive positions, in a positional notation, by the addition of some standard termination to the name of the number of the position. In terms of the wires of an abacus, this notation may be described as follows. The first wire we leave unnamed, the next is called the *one-ard*, then comes the *two-ard*, then the *three-ard*, and so on. In this notation a million is a six-ard, and a trillion is a

(two one-ards and four)-ard.

Whatever device is adopted the representation of very large numbers is exceedingly cumbrous. Refinements of notation increase the number of numbers susceptible of a brief representation, but inevitably numbers are reached too great for single-fold expression.

Important though the invention of a scale of notation has been in the development of arithmetic, this device tends to obscure the fundamental character of the number concept, for the kernel of the concept lies in the primitive tally. Let us suppose we make a primitive tally in marks on a strip of paper, for instance, for a flock of a dozen sheep, thus:

I   I   I   I   I   I   I   I   I   I   I   I

The notation is adequate provided the sign complex is regarded as a whole. Ambiguity arises, however, if we ask ourselves whether this tally records a *single* flock or not. There is nothing in this form to distinguish between the tally of a flock of a dozen sheep and the tally of, say, one half dozen, placed alongside the tally of another half dozen. From this point of view, the notation is not sufficiently closely knit, fails to delimit the bounds of the sign and fails to combine its parts into a single whole. To cement the several parts of the sign, one to another, we introduce a symbolic abbreviation for the word 'and', which we write '+' (the sign itself is a corruption of the initial letter of the Latin *plus*), and insert it between each pair of consecutive marks. The resulting sign takes the form

$$I + I + I + I + I + I + I + I + I + I + I,$$

which is now free from ambiguity. The sign is read as 'one and one and one . . . and one'. However, another objection arises now, namely that the sign 'I' is being used in two different ways. Each 'I' after the first is prefixed by the sign '+', but the first 'I' has no prefix. If we simply write another '+' in front of the first 'I', the whole value of using the '+' is lost, since the sign complex is again rendered ambiguous (compare $+ I + I + I + I + I$ with $+ I + I \quad + I + I + I$). The simplest way to maintain the force of the + sign, and to limit the use of the unit sign to the single '+ I' is to start the row of units with a non-significant commencement sign, a sign which by itself denotes, not a tally of a flock, but an absence of objects, like an abacus wire empty of beads, or an empty sheep pen. The zero, which we have already used to denote an empty wire, now serves also as a commencement sign for a tally, and the tally of a dozen becomes

$$0 + I + I + I + I + I + I + I + I + I + I + I.$$

This sign is constructed entirely from the commencement sign 'o' and the iteration of '+ I', its beginning and end are clearly delimited and its several parts are united in a comprehensive whole. If we characterise such a sign, for the moment, as a numeral pattern, then we perceive that numeral patterns have — that property of being contained one inside another to which we have already drawn attention in relation to dot patterns and prototype stick patterns. This is exhibited in the following diagram by the use of brackets:

$$((((((o + 1) + 1) + 1) + 1) + 1) + 1).$$

Each sign complex between brackets is (ignoring the interior brackets) a numeral pattern. From any numeral pattern another is constructed by affixing a terminal ' + 1', and accordingly the process of numeral pattern construction is *without end*.

To formulate a general definition of numeral patterns we introduce some arbitrary letter, x (or whatever distinguishing mark we please), and define the numeral patterns as the sign complexes obtained from the sign 'x' after eliminating 'x' by either the operation of replacing 'x' by '$x + 1$' or replacing 'x' by 'o'. Thus, for instance, the pattern $o + 1 + 1 + 1$ is obtained after four operations, deriving in turn x, $x + 1$, $x + 1 + 1$, $x + 1 + 1 + 1$, $o + 1 + 1 + 1$; here we replace 'x' by '$x + 1$' three times in succession, and then replace 'x' by 'o'. This definition in effect says that a numeral pattern is either 'o' or the sign obtained by affixing ' + 1' to a numeral pattern. The letter x introduced above is a symbol for a numeral pattern, and is called a *numeral variable*. The usage of a numeral variable x is determined by the rule that 'x' may be replaced by 'o' or '$x + 1$'.

The transformation of numeral patterns into conventional numerals is effected by a process akin to counting. We define in turn the four numerals 1, 2, 3, 4, etc., by the formal equivalences.

1 is $o + 1$, 2 is $1 + 1$, 3 is $2 + 1$, 4 is $3 + 1$, etc.; by means of these equivalents we may transform any numeral pattern into a conventional numeral; for instance

	$o + 1 + 1 + 1 + 1 + 1 + 1$
transforms first into	$1 + 1 + 1 + 1 + 1 + 1$
then to	$2 + 1 + 1 + 1 + 1$
and then in turn to	$3 + 1 + 1 + 1$
	$4 + 1 + 1$
	$5 + 1$
	$6 + 1$
	7

The transformation process amounts to enclosing the successive numeral patterns, contained in the composite pattern, between brackets, and then starting from the innermost bracket, replacing, in turn, $0+1$ by 1, $1+1$ by 2, $2+1$ by 3, and so on. This is exhibited in the following diagram, by deleting in turn the pattern replaced.

$$\overset{\cancel{1}\quad\cancel{2}\quad\cancel{3}\quad\cancel{4}\quad\cancel{5}\quad\cancel{6}\quad 7}{((((((\cancel{0}+\cancel{1})+\cancel{1})+\cancel{1})+\cancel{1})+\cancel{1})+\cancel{1})+\cancel{1}).}$$

A better illustration could be given in sand, by rubbing out each pattern when it is replaced by another. The point we are seeking to emphasise is that the transformation process is of a serial character, and operates like an automatic counting machine. The repeated copying of the 'tail' of the numeral pattern, shown in the first diagram, is no part of the process itself, and is a feature only of the method of illustration.

Using the familiar equality sign '=' between symbols to denote their equivalence, that is to say, to express the fact that one symbol may be replaced by the other, the foregoing equivalences are expressed by the formulae

$$1 = 0+1,\ 2 = 1+1,\ 3 = 2+1,\ 4 = 3+1,\ \text{etc.}$$

These formulae are the foundation of arithmetic, and the *conventional* use of number-signs which they exemplify is typical of the whole of arithmetic. The formulae are valid in the same sense that a dictionary is valid, each entry in the dictionary expressing the equivalence of two words, or phrases, this equivalence being established only by custom and convention. The use of number signs in language, as opposed to the use in arithmetic, is of an entirely different character; is I say that there are three matches in my pocket, then what I say is true or false, not solely in virtue of the conventions that $1+1 = 2$, $2+1 = 3$, but in virtue of what matches I actually have in my pocket. A fact, not a convention, is now in question. That it is the same thing to say I have three matches, as to say I have two and one matches is a consequence of the *convention* $2+1 = 3$, but this does not decide the truth of the proposition, whichever sentence it is expressed by. The question is simply 'Do the matches in my pocket form a three-sign or not?' And this is the same as asking

E

such a question as 'Is there a question mark on this page?' And the test for this is the empirical one of observation. No convention of arithmetic is a substitute for experiment.

To emphasise the correspondence between counting and transforming numeral patterns, consider the question 'How many a's are there in the following line?'

$$a \quad a \quad a \quad a \quad a \quad a \quad a$$

If we recall the steps by which the numbering of a flock is effected, we see that two processes are involved. First the group to be counted is transformed into a numeral pattern by *substituting* for each element of the group the sign $+1$, and then the resulting numeral pattern is transformed into a conventional numeral (or of course we may employ dot or stick patterns instead). Both these processes are illustrated in the following diagram, which gives a schematic representation of the process of counting the line of a's:

$$\mathit{1} \quad \mathit{2} \quad \mathit{3} \quad \mathit{4} \quad \mathit{5} \quad \mathit{6} \quad 7$$
$$\emptyset + \mathit{1} + \mathit{1} + \mathit{1} + \mathit{1} + \mathit{1} + \mathit{1} + \mathit{1}$$
$$\mathit{a} \quad \mathit{a} \quad \mathit{a} \quad \mathit{a} \quad \mathit{a} \quad \mathit{a} \quad \mathit{a}$$

We have indicated the substitution of one sign for another by deleting the sign which has been replaced. The operations are, carried out from left to right, new signs being written above instead of in place of, the signs they replace.

Comparing this diagram with the foregoing it is evident that the only difference between counting objects and transforming a numeral pattern lies in the initial step of replacing each object by '$+1$'. Making a tally, as opposed to counting, stops short at this initial step. We shall call the process of replacing each object of a group by the sign '$+1$' (actually or just in imagination) *regarding the group as a number sign*, so that we may say that making a tally consists only in regarding a group as a number sign, and counting consists in transforming a group, regarded as a number sign, into a conventional numeral. *Counting is not a process of discovery, but of transformation.*

Except in the case of bad handwriting or imperfect printing, the question whether, for instance, a certain sign is a letter a or not, simply does not arise. There is no unique pattern to which a sign must conform to be a letter a, but the accepted range of variation of signs which are recognised as a's is quite small, and throughout the range of variation the signs are recognisably different from any in the range of variation of another letter. In fact, although there is no standard a comparable with the standard yard, the various a signs may be regarded as copies of a prototype a, just as the signs which a child makes are copies of those which his teacher sets before him.

This matter of copying a sign is of fundamental importance, in arithmetic as in language. The relationship between a copy and that of which it is a copy is by no means an obvious one. Perhaps the simplest and most direct copy is a *tracing*, for what is called a free-hand copy, involving as it does the coordination of hand and eye, leaves open the question of the *method* of copying; but even a free-hand copy may be matched against the original, perhaps by superposition, or even by means of a double tracing, so the test of the copy, as a *copy*, may be held to lie in this matching, independently of the way in which the copy is obtained. In this sense a poem recited by heart is a copy of the poem learnt, despite the unknown mechanism which connects the two, and so too a written letter a is a copy even though there is no a, or image of an a, before us when we write it. It cannot be essential that, when we write a letter or recite a poem, we have an image before our mind's eye, for such an image could only be the product of some habit of the imagination and therefore of the same kind as a habit of the hand or tongue. This shows that what constitutes a copy is not the method by which it is obtained, but the test by which the copy is matched against the original. (We are not here concerned with such questions as the infringement of copyright, where the *method* of copying is the very point at issue.)

In arithmetic the problem of checking a copy is two-fold. First there is the matching of simple signs, like one '3' with another '3', and this is exactly the same as matching one letter with another. But in addition we have the more elaborate problem of matching two complexes of signs, and this may often be done in two different ways. Consider for example these two numeral patterns:

$$0 + 1 + 1 + 1 + 1 + 1 + 1 + 1 \qquad 0 + 1 + 1 + 1 + 1 + 1 + 1 + 1.$$

Are these the same pattern or not? If we transform each to a conventional numeral, we obtain the same numeral each time, namely 7. Alternatively we may match the signs one against the other, either by direct superposition, or by crossing off from the left, or even by conjoining the signs by a series of lines. It does not matter which method of matching we accept, so long as *some* method of deciding whether or not the signs are the same is laid down. Different methods of matching may not give the same result (even such methods as we have just described) and this possibility gives rise to an interesting question which we shall consider shortly.

Making a tally, which we have called regarding a group as a number sign, is the same operation as copying, or matching, a complex sign.

We have followed the development of the notations of arithmetic from the primitive tally stick to the cursive numerals of current usage, and we turn now to consider the antithesis expressed by the words *numeral* and *number*.

The terms *numeral* and *number-sign* are synonymous, so that the antithesis under consideration is between 'number' and 'number-sign'. Number-signs are not simply what we have called conventional numerals, but any of the signs we have described from a simple tally stick to a dot or stick pattern complex in positional notation. But if every sign we have so far described is a number-sign, what are the numbers of which these are *signs*? We answered the question 'What is the number of sheep in this flock?' by constructing a number sign. Does this mean that number is just another word for numeral and that *number-sign* is a nonsensical term? Certainly there appears to be nothing apart from the numerals to which we can point when we try to answer the question 'What is a number?' Words like 'table' and 'chair' are tokens which we can exchange for actual tables and chairs, but the word 'number', looked at in this way, seems to be a base currency. Yet if we try to identify numeral and number we meet an insuperable difficulty in the multiplicity of number signs. Which of the many three-signs, for instance, is the number three? Is it 1 1 1 or 0 + 1 + 1 + 1 or ∴ or 3? And why not x, y, z or a trio of sheep? Of course we may seek for a way out of this difficulty by saying that the number three is not any one of these signs, but is the totality of them. Superficially this answer seems to be adequate, but further consideration tends to weaken its force. When we talk of the totality of three-signs, presumably we mean not only the totality of three-signs hitherto written down or perceived, but must include also every future instance of a three and every symbolic representation of a three that may be conceived of. Thus the totality is not some definite group of things, something complete in itself; it must therefore be specified, not by an enumeration, but by the *property* which its members have in common. But if we identify a number with a totality defined only by the common property of its members, then we are in effect saying, not that a number is a totality of number-signs, but that number *is* this common property. And the only thing which, for example, the three-signs have in common, is their *common usage* in arithmetic as copies or transforms of the numeral pattern $0 + 1 + 1 + 1$.

Accordingly, a number is not a numeral, nor a collection of numerals, but is the role played in arithmetic by a numeral, that is, the role taken by a sign when it is used as a copy of a transform of a numeral pattern.

An analogy may help to make this clearer. Suppose we ask 'Who is Shakespeare's Hamlet?' (assuming, as is in fact the case, that Hamlet is not the name of a former Prince of Denmark). As with a number and the signs that denote it, Hamlet is not the actor who plays the part, nor the totality of such actors, but what they all have in common, namely the *role* which they are playing, and it is this role which is Hamlet.

Number signs have many uses in language, though their use in counting is perhaps the most important. Next to counting, the commonest service they perform is the representation of *rank* or *order*, as for instance when we speak of the *fourth* day of the month or the *seventy-first* year of a man's life. The process of numeral pattern construction is a *serial* process in which each pattern has by definition a successor, the successor of any pattern x being the pattern $x + 1$. Starting with the pattern 0 we construct *in turn* the patterns $0 + 1$, $0 + 1 + 1$, $0 + 1 + 1 + 1$, and so on. Each sign constructed is a *proper part* of its successor, so that the process never leads back to a sign previously constructed, and this is what we mean when we say that there is an *infinity* of numbers (the term infinity denoting simply *without end*). It may well be that in *practice* there is a limit to the numbers we may write down, since, for instance, the very mass in ink and paper of the numbers constructed, if the process were carried on long enough, would eventually exceed the mass of the earth and destroy every particle of life upon it. All that is implied in saying that there is an infinity of numbers is that no limit is laid down to the repetition of the number construction process. It is the *possibility* of constructing numeral patterns that is without end, not the actual construction process.

Since the numeral patterns do not form a closed cycle, no numeral is amongst its own successors, and so numeral patterns (or their abbreviations) supply an endless source of labels by which we can distinguish the elements of a group however great. For comparatively small groups, alphabet letters serve as well as numerals for labels, but the alphabet letters have an established order only in virtue of some accepted standard list, whereas the order of the numerals derives from the very process of generation, and if the numerals are expressed in a sufficiently condensed notation this order is immediately perceptible. (It is almost certain, moreover, that our established order for the letters of an alphabet is a much later innovation than ordering by numbers.)

The difference between counting and ordering by numbers, from the standpoint of stick patterns, is that in counting the stick patterns are *transformed* at each stage into the succeeding pattern, no record of the pattern transformed being kept, the final pattern obtained representing the number counted, whereas in ordering by numbers each pattern constructed is retained and is called the *ordinal* number of the last object counted. To take a concrete example, if we count the rows of hoops o o o o by stick patterns all that we retain is the single pattern □, but if we seek to order the hoops we obtain in turn

$$\text{I} \qquad \text{I L} \qquad \text{I L L} \qquad \text{I L L □}$$
$$\text{ø o o o,} \quad \text{ø ø o o,} \quad \text{ø ø ø o,} \quad \text{ø ø ø ø}$$

and the final configuration is retained *in toto*, each stick pattern representing the ordinal number of the hoop beneath it. In terms of conventional numerals the difference is between

$$1 \quad 2 \quad 3 \quad 4 \qquad 1 \quad 2 \quad 3 \quad 4$$
$$\text{and}$$
$$\emptyset \quad \emptyset \quad \emptyset \quad \emptyset \qquad \emptyset \quad \emptyset \quad \emptyset \quad \emptyset$$

which brings to light the important fact that the very process of counting establishes the objects counted in (serial) order. In counting we *ignore* (that is, overlook) the ordering by numbers which takes place on the way, and so there is a sense in which the ordinal use of numerals is more fundamental than their use in counting; in spite of this the number of a group is called a *cardinal* number.

The cardinal number of a group represents the group *as a whole*, but ordinal numbers serve to *distinguish the elements*. When a line of soldiers 'numbers off', each soldier calls his ordinal number and the ordinal number of the last soldier is also the cardinal number of the line. Moreover the ordinal number of any soldier is also the cardinal number of the section of the line which that soldier terminates. Thus the *cardinal* number of a group is also the ordinal number of an element of the group and is amongst the successors of the ordinal number of every other element.

Another way in which we can bring out the difference between cardinal and ordinal numbers is by observing that the ordinal number of an element X of a group is the successor of the cardinal of the subgroup of all the elements counted before X. Thus a man is ninth in a line if he follows a group of eight men. Observe that the man's ordinal number is independent of the actual order of the men in front of him (and of the men after him), and depends only upon the *cardinal* number of the group that precedes him. The terms 'precede' and 'succeed' which we have used in relation to the elements of a group refer to the order established by the process of counting or making a tally, elements already counted at some stage in the process being said to precede the elements not yet counted, and the latter being said to follow or succeed the former. We have already observed that counting necessarily establishes an order among the objects counted, namely the order of selection. To count a field of sheep the sheep must be selected one after the other, and some device adopted for distinguishing those already counted from the rest;

this may be done by passing the flock through a gate, or marking each one counted. We could also label each sheep with its ordinal, that is, the number which succeeds that on the last label attached, or, what is the same thing, the number which succeeds the cardinal of the group so far counted. But if our object is to guard against loss from the flock the use of ordinal number is no added security, for even if we find sheep number eleven, sheep number three may be missing.

The elements of a (finite) group may be said to have an *order* if, and only if, each element, save one alone, has a unique successor, and each, save one alone, a unique predecessor. In such a case as men in a line the relations of *left hand* and *right hand* serve to determine each man's successor and predecessor, but for a general group the determination of predecessors and successors involves an *ad hoc* act of choice. If it is (physically) impossible to make such a choice (as, for instance, when the elements of a group are in rapid motion) then ordering is impossible—and so, too, is counting. The elements of a group may or may not admit a unique order in relation to some attribute of the elements. For instance, a line of men may be arranged in descending order of heights provided no two are of the same height, but if there are two of equal height, the ordering is not complete. It is important to remember that such an ordering (when it is unique) may be attained without the use of numbers. It is not at all necessary that the various heights be expressible in multiples of some unit; provided we can say of any two men of the group which is the taller (by placing them back to back, for instance), the whole line may be arranged in descending order of heights. We start with any two, and settle their order. A third is then put in relation to the first two, simply by comparing him with each of them (he will stand in front if taller than the first, at the back if the second is taller than he, and between the two otherwise). A fourth is then brought into place, by fixing his position with regard to each of the first three, and so on.

The words *first, second, third, fourth*, and so on, stand for the attributes *numbered one, numbered two, numbered three, numbered four*, etc., that is to say, the first day is that which bears the number one, the second house is that which bears the number two, and so on. By 'bears the number', however, we mean a variety of things. The house *bears* its number on the door, a day *bears* its date number in the calendar, and the third soldier in the line *bears* the number three only in the sense that he called himself *three* when the line numbered off. When we say that, for instance, seven is the seventh number, we say no more than that the number seven is the number seven. This is bound up with the common fallacy that it is significant to say that there are six numbers from one to six; for if we counted the numbers one to six we should merely write down (or say) the numbers one to six twice over. The numbers one to six are, so to speak, the standard by which we judge a group to have six members, and it no more makes sense to say that there are six numbers from one to six than to say that the standard yard is one yard long. If we see a row of houses numbered from one to six, then the row has *already been counted* for us.

In focusing attention on the need to order the elements of a group which is being counted, we bring to mind the question of the dependence of the cardinal number of a group upon the order of its elements. How do we know that changing the order will not change the cardinal of the group? This is, in fact, no more than a part of the wider question: if we count the elements of a group more than once (in the same order or not) will we necessarily obtain the same cardinal each time? Certainly, as we all know, counting and recounting large groups often leads to conflicting results, and the usual supposition is that in some of the counts we have made a mistake, and that there could really be only one correct answer. The problem, however, is deeper than this leads one to suppose. Sometimes we do find that we have made a mistake; we *remember* saying 'seven and one is nine' at some point in the counting, or we notice an element previously overlooked, or an element counted twice. But if we *cannot* find a mistake, are we forced to concede a mistake entirely on the strength of the divergence of the answers? In practice we deal with a situation of this kind in a variety of ways. We establish various precautions to guard against error, such as having independent witnesses to watch that no element is added or removed during the counting, and arranging for several people to take part in the recounts. Men are classified as good or bad counters according to the frequency with which they obtain the same answer, and so on. Yet it might happen that, despite every precaution, in counting and recounting a particular group we continue to obtain a variety of answers, or perhaps two answers at alternate counts, like six million and three alternating with six million and four. Such a group might of course be dismissed as *uncountable*, and various hypotheses formulated to account for the mystery (such as elements coalescing and separating on alternate counts). Or we might accept the disjunction *six million and three or four* as the answer, or to select the lesser or greater according to circumstances. Another possibility would be to accept the two answers as evidence of two facts, like a drawing which may be looked at in two ways (the so-called optical illusion). To say that there *must* be some one definite answer, whether we can find it by counting or not, says nothing at all, for, by definition, the number of a group is found by counting.

A group yielding different cardinals on successive counts is, to draw an extreme comparison, like a stick pattern changing from four to five before our eyes despite every precaution against human interference or fraud. If such occurrences were the commonplace, numbers would entirely lose their practical value.

We are so familiar with the use of numbers in language, in our everyday life, that we are too readily apt to suppose that we could not dispense with them without an intolerable sacrifice of our means to describe the world as we see it. There is no doubt that without numbers language would be exceedingly cumbrous and complicated, but a language lacking number words can nevertheless express everything that can be said in the current medium of expression. That is the meaning of the analysis of number which we have given above. Numerals are *abbreviations* and every abbreviation is dispensable. The route which leads from a group to its cardinal (or to the ordinal of a term of the group) is traversable *both ways* and everything we express by means of numbers may be expanded in sentences from which the number words have been eliminated. Since *a car and a car and a car* is three cars, the sentence 'I bought three cars this year' says only that 'I bought a car and a car and a car this year', and there are no number words in this latter sentence. You may say if you like that *a car and a car and a car* is an instance of the number three, but this simply means that if you establish an arithmetic *then* you regain your number words; we are under no necessity to do so, other than the urge for brevity. From this point of view one may in fact doubt that arithmetic has any value. If 'three cars' is just another way of saying 'a car and a car and a car', why do we bother to introduce this new term *three*, or, to use a mathematical expression, why do we introduce a *new notation*? And the answer is that new notations are amongst the finest and most valuable discoveries of which the human mind is capable, for a notation is something which does our work for us. Try and multiply three hundred and sixty-four by six hundred and seventy-two, first on an abacus and then in our modern notation and see how much work the notation does for you.

The elimination of number words, whilst necessarily possible, is not always so straighforward as the previous example suggests, owing to the maturity and flexibility of our language and to the vast range of services which numbers perform. Ordinals, how-ever, present little more difficulty than cardinals. If I say, for instance, that I bought my third car this year, I can eliminate the ordinal *third* by saying instead that I bought a car after I had bought a car and a car this year.

The *adverbial* use of number exemplified by the words once, twice, thrice, etc., is equally easy to understand provided we remember that adverbs qualify phrases and not just nouns. Thus 'he paid his fare twice' means 'he paid his fare and he paid his fare' and *not* 'he paid his fare and his fare' (which is 'he paid two fares').

The terms single, double, treble, etc., are used both as adjec-tives and verbs. The adjectival use, is, with certain exceptions, purely a literary variant of the ordinary cardinals. To pay double fare is to pay two fares, and a double-deck bus is a two-deck bus. But a double bed is one bed for two persons. 'Double' used as a verb is often particularly difficult to eliminate. 'Care-ful driving doubles the life of a tyre' may perhaps be rendered 'Careful driving makes a tyre last a life and a life', where we have first transformed the verb 'to double' into the adjective '*double*'.

V

THE DEFINITION OF NUMBER

IT is surely a very remarkable thing that despite the range, power and success of modern mathematics, the concept of natural number, on which the whole edifice rests, is still something of a mystery.

Towards the end of the last century the great German mathematician and philosopher Gottlob Frege, dissenting strongly from the view current in his time—and perhaps still today—that number is either a physical or a mental entity, gave the first purely logical definition of a natural number. His work aroused so little interest at the time that his definition remained practically unknown until it was rediscovered by Bertrand Russell 20 years later. The definition, though a little strange at first sight, is quite simple and is based upon a notion which is of first importance in many branches of mathematics, the notion of a $(1, 1)$ correspondence.

Frege defined number in terms of the concept 'equinumerous'. That is to say, he sought first to define the property of 'having the same number', and then to define the concept of number itself by means of this property. The reasoning which led Frege, and Russell after him, to pursue this course seems quite plain. Just as Plato sought to define the predicate red as that which all red objects have in common so Frege identified each natural number with the property common to all the classes which have this number of members. Of course, if we are to avoid a vicious circle in the definition we must define the property of having the same number of terms without reference to the actual number, and this is where the concept of $(1, 1)$ correlation plays a decisive part. If there is a collection of cups and saucers on the table, and if there is a cup on each saucer and a saucer under each cup, then we can say that the class of cups and the class of saucers have the *same number* of members, even if we do not know what this number is. Provided, therefore, that we can define $(1, 1)$ correlation independently of the number concept then by its means we can define the property of having the same number of terms without any reference to numbers. Leaving on one side for the moment the problem of defining (1, 1) correspondence the next stage in the Frege-Russell definition of number is the identification of a natural number n with the property of being (1, 1) related to a class of natural number n; if we make no distinction between a class and the condition for membership of the class this amounts to identifying n with the class of all classes (1, 1) related to some class of n members. There remains the task of providing sample classes for each natural number. For the number zero, we take the empty class (which is defined by the property 'not identical with itself'); then as a *sample* class of one member we take the class (0) whose sole member is the number zero; for the sample class of two members we take the class whose members are 0 and 1, and so on. A class is said to have 2 members for instance if it is a member of the class of classes 2, that is to say if it has the property of being (1, 1) related to the class (0, 1).

To complete the Frege-Russell definition of number we have to show that the concept of a (1, 1) relation is definable without using the notion of number. Let us first take the idea of *relation* for granted and concentrate upon the specific property of one to one relatedness.

Denoting by x, y z variables for *things*, a generic term for the objects of our universe of discourse, and by xRy the statement that x stands in the relation R to y, the conditions for a relation R to be a (1, 1) relation are

$$xRy \ \& \ xRz \rightarrow y \equiv z, \quad yRx \ \& \ zRx \rightarrow y \equiv z$$

As to relation itself, a relation may be defined simply as a class of ordered pairs, so that a relation R is the class of all ordered pairs (x, y) such that xRy. Finally, we may define an *ordered* pair (x, y) as the class whose only members are x, and the class whose members are x and y, so that the ordered pair is a class of two elements, one of which is distinguished by being a member of the other.

In this way the several elements in the Frege-Russell definition have all been defined in terms of the single concept of class, and there the reduction process rests with class as the sole concept.

Apart from the logical connectives *and, or, not, implies,* the only connectives introduced in the definition of number are class membership and identity, and in fact the second of these may be defined in terms of the first.

For instance we may take

$$(a \equiv b) \longleftrightarrow (\forall x)(a \in x \longleftrightarrow b \in x)$$

where '$\longleftrightarrow$' is the logical connective 'implies and is implied by', ($\forall x$) is a prefix denoting universality, and '$a \in x$' reads 'a is a member of x'. The definition of number involves therefore only a single concept and a single binary connective.

The Frege-Russell definition, like a great part of modern mathematics, can only be expressed in a logic capable of making universal statements about classes. Class logic, as we may call it, has however the doubtful distinction of being perhaps the most spectacularly unfortunate of man's creations. Frege's system of class logic was found to be self-contradictory just as his second volume was passing through the press. It was a bitter blow to a man who had received little enough encouragement from his contemporaries, and as he himself observed, small consolation to him that the contradiction was not peculiar to his system but was common to all work on set theory current in his time. The contradiction arises when we consider the class α whose members are those classes which are not members of themselves. By definition, x belongs to α if and only if x is not its own member, i.e.

$$(x \in \alpha) \longleftrightarrow x \notin x.$$

Taking α for x we obtain the blank contradiction

$$(\alpha \in \alpha) \longleftrightarrow (\alpha \notin \alpha).$$

Many unsuccessful attempts have been made to formulate a demonstrably consistent class logic. Russell's system has the safeguard of a hierarchy of types which places a class in a type above that of its members and bans the formation of classes whose members do not all belong to the same type; one of the disadvantages of this system is that it has to postulate the existence of an infinite set to ensure that there is no greatest number. Another is that it furnishes us with not one sort of natural number, but infinitely many sorts, one in each type. A system of

Quine's avoided this reduplication of the natural numbers at the cost of making it impossible to prove there are n numbers from 1 to n; this system has recently been shown to contradict the axiom of choice. Another system, due to Zermelo seeks to keep out of trouble by restricting the membership of new classes to members of existing classes, making the generation of new classes rather uncertain and laborious. Von Neumann and Bernays have formulated systems in which certain classes only are distinguished as elements and only these distinguished classes are allowed to be members of classes. A later system of Quine which also introduced a condition of element-hood was shown to be self-contradictory on the eve of publication. Quine's system was repaired by Hao Wang who has just recently produced a set theory of his own which admits a transfinite series of levels.

In view of the dangers to which class logic is exposed it is obviously desirable to find a treatment of the natural numbers which can be formulated in a more economical and relatively safer level of logic. Of course if the concept of class was intrinsic to every branch of mathematics there would be no point in seeking for a class-free definition of number, but in fact there are interesting and important branches of mathematics, for instance the classical theory of numbers and axiomatic projective geometry, in which the class concept is totally dispensable.

It is actually quite easy, following Peano, to give an account of arithmetic in which number is taken as a primitive concept. If we denote number variables by a, b, c and take S, 0, = as primitive constants (where 'S' may be interpreted as 'successor of') then arithmetic may be based on the axioms

$$a = b \longleftrightarrow Sa = Sb, \quad Sa \neq 0$$
$$(a = b) \rightarrow ((a = c) \rightarrow (b = c))$$
$$a + 0 = a, \quad a + Sb = S(a + b)$$
$$a \cdot 0 = 0, \quad a \cdot Sb = a \cdot b + a$$

and the induction axiom

$$\{P0 \ \& \ (\forall a)(Pa \rightarrow PSa)\} \rightarrow Pa$$

With the aid of a suitable system of logic (quantification theory with number variables only) these axioms suffice for the proof of all the familiar theorems of elementary arithmetic. We can in fact dispense with logic entirely (and with the first three axioms) if we make provision for the definition of new functions. However the question in which we are now interested is the definition of number. To what extent can these axioms be said to constitute a definition of number?

What the axioms in fact provide are transformation rules for certain patterns. For instance they enable us to pass from the pattern $So + SSo$ in turn through $S(So + So)$, $SS(So + o)$ to $SSSo$. If we now introduce the familiar abbreviations $1 = So$, $2 = 1 + 1$, $3 = 2 + 1$ and so on, then we are able to make the passage from $1 + 2$ to 3. But what has this to do with numbers? Are numbers no more than shapes, like '2' or '5'. This is obviously not the case since numbers have such properties as being odd or even, whereas patterns like '2' or '5' are printed in ink, or painted on the gate. Are there perhaps no such things as numbers, only numerals?

Instead of pursuing the question whether numbers *exist* let us turn for a moment from mathematics to chess to consider the existence of the king of chess. Shall we say that the king of chess exists or not? Obviously the king of chess is not simply the particular piece of wood or ivory which is moved about on the chess board. For one thing the game could be played with a cork in place of the king, or for that matter the king and queen pieces could be interchanged. Now what does interchanging the king and queen consist in? It does not involve any alteration to the pieces themselves, but only to the moves which they make. It follows that what constitutes a piece king are not its shape or size but the kind of moves it makes. A piece is king because it makes the king's moves in the game. Thus we can say that the king of chess is one of the parts which a piece plays in the game of chess, and we can characterise this part by describing the moves.

The king of chess is a role in a game in precisely the way that the King of Ruritania is a role in a play. And so too the natural numbers are roles in arithmetic, the number two for instance being the role which arithmetic assigns to the numeral 2, and so on.

Russell and Frege both opposed the use of number variables as primitive signs on the grounds that numbers belong, not only to arithmetic but also to everyday life. Thus we use numbers not only in such contexts as $2+3=5$ but in shopping and counting planets. On the basis of the Frege-Russell definition we can prove in class logic such assertions as:

'If A, B, C are white then three things are white,'

and such statements as this were held to be incapable of proof if numbers were taken as primitive symbols. However, nothing more is needed than the introduction into arithmetic of a suitable counting operator. If x is a variable for names A, B, C, . . . and ξ is a variable for classes of names like A & B & C, then we may define a counting operator N by the recursive equations

$$N(x) = 1$$
$$N(\xi \ \& \ x) = N(\xi) + 1,$$

From these equations we may derive in turn

$$N(A) = 1, \ N(A \ \& \ B) = N(A) + 1 = 1 + 1 = 2,$$
$$N(A \ \& \ B \ \& \ C) = N(A \ \& \ B) + 1 = 2 + 1 = 3, \quad \text{and so on.}$$

Although formalised arithmetic is adequate both for mathematics and logic it is an extraordinary fact that no formalisation of arithmetic is adequate to characterise the number concept completely. Every axiomatic system admits a valid interpretation in which a richer class of elements than the natural numbers plays the number role, and this is equally true whether we take numbers as primitive symbols or define them in class logic as in the Frege-Russell definition.

I am going to turn now from natural numbers to consider other number systems.

Integers

The passage from natural numbers to integers presents no difficulty. Probably the simplest course is to define an integer as an ordered pair of natural numbers $[a, b]$ with an arithmetic

$$[a, b] \gtreqless [c, d] \longleftrightarrow a + d \gtreqless b + c$$
$$[a, b] + [c, d] = [a+c, b+d], \quad [a, b] - [c, d] = [a+d, b+c],$$
$$[a, b] \ . \ [c, d] = [ac+bd, ad+bc].$$

It readily follows that $\quad \{[a, b] + [c, d]\} - [c, d] = [a, b]$
$$\{[a, b] - [c, d]\} + [c, d] = [a, b],$$

F

and, most important, that

$$\{[a,\, b] = [a',\, b'] \,\&\, [c,\, d] = [c',\, d']\}$$
$$\rightarrow\{[a,\, b] \pm [c,\, d] = [a',\, b'] \pm [c',\, d'] \,\&$$
$$[a,\, b] \cdot [c,\, d] = [a',\, b'] \cdot [c',\, d']\}.$$

If $a \geqslant b$, $[a,\, b] = [a-b,\, 0]$ and if $a < b$, $[a,\, b] = [0,\, b-a]$ so that every integer is expressible in one of the forms $[x,\, 0]$, $[0,\, x]$.

Writing $+x$ and $-x$ for $[x,\, 0]$, $[0,\, x]$ respectively we regain the familiar representation of integers by $+$ and $-$ signs. The so-called rules of sign for integers are of course provable consequences of the defining equations for addition, subtraction and multiplication. For instance the rule 'minus times minus makes plus' is proved by showing

$$(-x) \cdot (-y) = [0,\, x] \cdot [0,\, y] = [xy,\, 0] = +xy.$$

All this is of course well known. It is, however, a not uncommon fallacy to suppose that for $+x$ we must take, not $[x,\, 0]$ as we have done but the class of all pairs equal to $[x,\, 0]$, i.e. all the pairs $[x+n,\, n]$, presumably on the grounds that we must not identify an integer with one of its representations. Since a concept is in fact no more to be identified with the *class* of its representations than with any one of them this re-introduction of the logic of classes into the theory of integers is without point.

The arithmetic of fractions, like the arithmetic of integers, is simply an arithmetic of ordered pairs of natural numbers p/q. Of rather more interest is the arithmetic of pairs of *integers* ξ/η, $\eta \neq 0$, with the rules

$$\xi/\eta \gtreqless \xi'/\eta' \longleftrightarrow \xi\eta\eta'^2 \gtreqless \xi'\eta'\eta^2$$
$$\xi/\eta \pm \xi'/\eta' = (\xi\eta' \pm \xi'\eta)/\eta\eta'$$
$$(\xi/\eta) \cdot (\xi'/\eta') = \xi\xi'/\eta\eta'$$

and
$$(\xi/\eta) \div (\xi'/\eta') = \xi\eta'/\xi'\eta, \quad \xi' \neq 0.$$

If these rules are fully expressed in terms of natural numbers they become rather formidable. At the price of a little awkwardness over division a simpler theory is obtained by taking rational numbers as triples of natural numbers $(p,\, q)/r$, $r > 0$. For equality and inequality

$$(p,\, q)/r \gtreqless (p',\, q')/r' \longleftrightarrow (pr',\, qr') \gtreqless (p'r,\, q'r).$$

Multiplication has the expected form

$$(p, q)/r \times (p', q')/r' = (pp' + qq', pq' + p'q)/rr'$$

and for division we may take

$$p' \neq q' \to (p, q)/r \div (p', q')/r' = (p, q)/r \times (p'r', q'r')/(p' - q')^2.$$

Complex numbers

The familiar definition of complex numbers as ordered pairs of real numbers with postulated laws of equality, addition and multiplication, as set out for instance in the Mathematical Association's last Trigonometry Report, is logically sound, economical and effective. It is sometimes criticised however from the pedagogic standpoint for its failure to indicate its motivation. Certainly the multiplication law is a strange one and its connection with $\sqrt{-1}$ by no means immediately evident, but I think this difficulty is intrinsic since every logical development of complex arithmetic must either postulate a number whose square is -1 or assure the existence of such a number by some devious route.

One such route is to define complex numbers as the residue class, modulo $\xi^2 + 1$, of the field of real numbers extended by an indeterminate ξ. The residues modulo $\xi^2 + 1$ are the binomials $a + b\xi$ where a, b are real numbers, and ξ satisfies the equation $\xi^2 = -1$ in virtue of the congruence

$$\xi^2 \equiv -1 \,(\text{mod } \xi^2 + 1).$$

Taking residues modulo $\xi^2 + 1$ avoids postulating the existence of a number whose square is -1 but it is by no means self-evident to the uninitiated that the residue class $\text{mod}(\xi^2 + 1)$ is a number field.

There are however more serious obstacles in the path of this development of complex arithmetic. The extension of the field of real numbers by an indeterminate ξ is defined to be the ring of polynomials

$$a_0 + a_1\xi + a_2\xi^2 + \ldots + a_n\xi^n$$

where $a_0, a_1, \ldots, a_n$ are real numbers. This ring in turn may be defined as the class of all ordered sets of real numbers

$$(a_0, a_1, a_2, \ldots, a_n), n = 1, 2, 3, \ldots$$

with postulated equality, addition and multiplication laws

$(a_0, a_1, \ldots, a_n) = (b_0, b_1 \ldots, b_n)$, if and only if

$$a_r = b_r, \ r = 1, 2, \ldots, n,$$

$$(a_0, a_1, \ldots, a_n) + (b_0, b_1, \ldots, b_n) = (a_0 + b_0, \ldots, a_n + b_n)$$

$$(a_0, a_1, \ldots, a_n) \cdot (b_0, b_1, \ldots, b_n) =$$

$$(a_0 b_0, a_0 b_1 + a_1 b_0, a_0 b_2 + a_1 b_1 + a_2 b_0, \ldots).$$

If the polynomial ring is defined in this way the definition of complex numbers as the residue class modulo $(1, 1)$ can no longer be preferred to the definition in terms of ordered pairs, since it verges on the absurd to replace a definition in terms of ordered pairs by a definition in terms of ordered sets of arbitrary length.

The polynomial ring can however be defined without introducing the concept of an ordered set. We introduce two primitives $+$ and $.$, a polynomial ξ and variables for polynomials $a, b, c, \ldots$ and lay down the axioms

$$a + b = b + a \qquad (a + b) + c = a + (b + c)$$

$$a \cdot b = b \cdot a \qquad (a \cdot b) \cdot c = a \cdot (b \cdot c)$$

$$(a + b) \cdot c = a \cdot c + b \cdot c.$$

The concept 'polynomial of degree n' is defined inductively as follows: Real numbers are polynomials of degree zero; ξ is a polynomial of degree unity; if a and b are polynomials of degrees m and n respectively then $a + b$ is a polynomial of degree $\max(m, n)$ and $a \cdot b$ is a polynomial of degree $m + n$. For equality between polynomials we postulate the inference rule:

If α, β are polynomials of degree zero and if

$$\alpha + a\xi = \beta + b\xi$$

then

$$\alpha = \beta \text{ and } a = b.$$

By induction over the degree of the polynomial we can prove that every polynomial of the system is expressible in the form

$$a_0 + a_1 \xi + a_2 \xi^2 + a_3 \xi^3 + \ldots + a_n \xi^n$$

where ξ^n is an abbreviation for the product $\xi \cdot \xi \ldots \xi$ with n factors, showing that the elements of the axiom system are the polynomials of the ring. Apart from the equality rule, the axioms are all of the kind familiar in elementary arithmetic, and as a definition of the polynomial ring could hardly be simpler, but as an introduction to complex arithmetic it is a formidable journey to make to avoid the postulation of the multiplication law for ordered pairs.

Real numbers

In talking about complex numbers I have taken the real numbers for granted and it is to them that I must now turn. The logical foundation of the theory of real numbers presents formidable difficulties. The statement and proof of *Dedekind's Theorem* for instance requires of course a fully fledged class logic for its formalisation. Along with the complexities and perils of a class logic we have two further bitter draughts to swallow. The first is that *no* formal system is rich enough to *define* all real numbers, because all the statements of a formal system can be enumerated in a simple sequence, and therefore at most a denumerable infinity of real numbers can be *defined*. And, definition apart, Cantor's proof of the non-denumerability of the class of real numbers is now known to establish only *relative* non-denumerability. For every formalised theory of real numbers can be shown to have a denumerable model; that is to say there is an interpretation of the predicates of the system under which every true statement about classes of integers becomes a true statement about integers. Since the integers are denumerable it follows that the class of real numbers in the system is denumerable. Cantor's proof of non-denumerability is therefore a proof of the incompleteness of the formal system, a proof that the function which is known to enumerate the class of real numbers is undefinable in the system. Even without appeal to the known existence of a denumerable model the relativity of Cantor's proof is apparent. For the proof starts by saying: 'let a_1, a_2, $\ldots$ be an enumeration of the real numbers' and this statement of course has a concealed existential premiss 'If the class of real numbers is denumerable in our formalism'. The conclusion to be drawn from the familiar *reductio ad absurdum* proof is therefore that the class is not denumerable in our formalism. To pass from this to absolute non-denumerability we must know that our system of class logic is complete, i.e. that *all* enumerations are definable in it, and precisely *this* proves not to be the case.

If we are content to formalise only a constructive fragment of real number theory we can dispense with class logic entirely and use restricted predicate logic or simply free variable logic.

In free variable logic however real numbers lose most of their familiar properties. Let $r_1, r_2 \ldots$ be one of the familiar enumerations of the rationals, and let us, following Dedekind, define a recursive real number to be a recursive predicate $P(n)$ such that there is no greatest r_n for which $P(n)$ holds and if $r_m < r_n$ then $P(n)$ entails $P(m)$. Further, let us say that a recursive sequence of rationals $s(n)$ is *recursive* convergent if there is a recursive n_k so that

$$n \geqslant n_k \to |s(n) - s(n_k)| < 10^{-k}.$$

Next we define a recursive decimal to be $\overset{\infty}{\underset{r=0}{\Sigma}} f(n) \; 10^{-n}$ with $f(n)$ recursive and $0 \leqslant f(n) < 10$ for $n \geqslant 1$, and a recursive nest of intervals to be a pair of recursive sequences of rationals a_n, b_n such that

$$a_n \leqslant a_{n+1} < b_{n+1} \leqslant b_n$$

and
$$n \geqslant n_k \to b_n - a_n < 10^{-k}.$$

If in these definitions we take recursive functions to be *primitive* recursive functions, the functions of elementary arithmetic, then none of the expected connections holds. It can be shown that there are recursive convergent sequences which have no recursive real limit, recursive real numbers which are not recursive decimals and recursive nests which do not contain a recursive real number.

If we interpret recursive in the much wider sense of general recursive, and use predicate logic then *some* of the expected connections can be established, but we are as far away as ever from classical analysis. There are recursive real numbers which can neither be proved equal, nor proved unequal; recursive real numbers which cannot be proved rational nor proved irrational and monotonic increasing and bounded sequences which have no recursive limit. In fact the function theory of a constructive fragment of the class of real numbers is not significantly richer than the system of analysis which can be developed in a rational field.

VI

MATHEMATICAL SYSTEMS

THE primary difficulty to be faced in a comparative survey of finitism and formalism is that these two mathematical systems appear to lie in different planes, for it is often said that the finitist is interested in the *meaning* of mathematics whereas the formalist is not; so that, it would seem, to choose between them there is nought but personal preference to guide us. This mistaken view arises from a failure to distinguish two widely divergent usages of the term 'meaning'; that is to say, from a confusion of what may very roughly be described as 'the psychological accompaniments of' the usage of words with that usage itself. In the *Logical Syntax of Language*, Carnap expresses an analogous view, maintaining that, since the rules of a Calculus may be chosen quite arbitrarily, practical considerations alone afford a standard of judgement between the formalist and finitist systems. One might express this by saying: Finitism and formalism have the relation of draughts to chess. In this form the argument is not, of course, open to our previous criticism (though Carnap does not seem to have been aware of the distinction we made above, as is shown by his remark that the chessmen, the symbols of the chess calculus, have no meaning), but I hope to show in the course of this paper that this argument, too, is erroneous and conceals a number of confusions.

Both the formalist and finitist claim to effect, in some sense, a logical foundation of mathematics. I believe, however, that neither can substantiate this claim; formalist and finitist alike deem classical mathematics to be a tottering edifice, and whilst supposedly engaged in the task of strengthening and cleaning this edifice, each in fact sets up a new structure, on a new site, and built of new materials. To change the metaphor, it is as though, becoming confused about a game they have played a long time, they seek to dispel this confusion, the one by denying that many familiar pieces belong to the game at all, the other by giving the pieces many new and diverse extra moves.

And this without either of them finding the source of their confusion.

Brouwer's analysis of the 'infinite' in mathematics, in spite of the fact that he himself misunderstood it and thought that he had found a *fallacy* in mathematics (as apart from a widespread confusion in the current interpretation of mathematics), is of great importance to mathematical philosophy. It was, however, of little significance until it became illuminated by Wittgenstein's far-reaching work on the elimination of metaphysics in mathematics. (A recent paper[1] which purported to give some references to Wittgenstein's work seems to me to show a complete misunderstanding of it, and to give a very false impression —partly owing to an unfortunate play upon the words 'meaning' and 'meaningless'.) Brouwer's original observation may be expressed in some such way as this: let $p(n)$ be a numerical proposition which we can test for any assigned value of n (e.g. the proposition '$2n$ is a sum of two primes'). Suppose that we test the proposition for $n = 1, 2, 3$, and so on, up to, say, 1000, and find each time that $p(n)$ is true. It clearly may happen that, for however great a value of n we test the proposition, we do not find a value for which it is false. Use the proposition ($2n$ is a sum of two primes) to determine a decimal d, thus: the nth digit of the decimal d is zero if $2n$ is a sum of 2 primes, and is unity if $2n$ is not a sum of 2 primes. (Observe that the decimal is determinate, since for any assigned value of n we can calculate whether $2n$ is a sum of 2 primes or not; in less then $n^2/2$ operations, in fact.) Is the decimal d, so determined, commensurable with zero? This takes us straight to the crux of the formalist-finitist opposition. The finitist would say that here we have a case of failure of the 'tertium non datur', for we can neither affirm that d is equal to zero, nor deny it (d is by definition non-negative). The formalist, however, would say 'd *must* be either equal to zero or greater than zero, though we have no way of deciding the question'.

At first sight the difference seems to be of little importance; for both accept the primary fact that 'we have no way of deciding the question'. What, then, is the significance of the different ways in which they qualify their acceptance? The former says 'this is a case of failure of the "tertium non datur" ', the latter '*d must* be either equal to zero or greater than zero'. There would, in fact, be no point in these qualifications, and therefore no disagreement between the parties, were it not for the application of these remarks in the so-called 'reductio ad absurdum proof' in mathematics, which in the present case would take the form: To prove *d* is equal to zero, we show that '$d > 0$' leads to a contradiction; the proof being completed by the addition of:—'but *d* is either zero or greater than zero and so *d* is equal to zero'. Brouwer's objection to this line of proof had its emphasis on this last step, 'either zero or greater than zero'. He considered that we must take account of a 'third possibility', viz., $d \geqslant 0$, but neither zero nor greater than zero (the idea being possibly suggested to him by the well-known 'problem' in transfinite cardinals as to whether one of the possibilities $c > \aleph_1$, $c = \aleph_1$, $c < \aleph_1$ must necessarily hold—it seemed that we might find that none of the possibilities held). The formalist's attitude to this objection is as follows. It is true, he says, that in the development of the decimal *d*, we cannot demonstrate either that all the digits are zero, or that there *is* a digit which is not zero, but we shall *postulate* that the 'tertium non datur' applies in this case, and say that one or other of the possibilities *must* occur, even though we don't know which. Considered superficially, the formalist thesis is quite in the spirit of traditional mathematics; axioms have long been a familiar part of mathematics. The formalist himself, however, is not fully satisfied by just laying down an axiom, but wishes to show further that his axioms are non-contradictory; ultimately he 'proves' that this cannot be demonstrated.[2] We shall for the moment leave on one side consideration of this 'proof', and return to the finitist's claims. What, in fact, Brouwer has accomplished is this: He constructs a word-series '*d* is equal to zero' which does not satisfy one of the *defining characteristics of propositions*, viz., $p \vee \sim p$; for one can neither assert '*d* is equal to zero' nor '*d* is not equal to zero'. His conclusion then should have been: *either* we change our definition of proposition and reject the defining characteristic $p \vee \sim p$, *or* we retain this characteristic and perceive that according to it the word series '*d* is equal to zero' is *no proposition*. Brouwer, in fact, chose the first alternative, but appears to have been aware of the second, since he referred to such word-series as meaningless. That this idea was misunderstood and disputed is much to be expected, for the sentential form 'd is equal to zero' leads us irresistibly to call it a proposition. It is this very temptation that has caused so much confused thought about the theory of types, and is at the root of the fascination which the Antinomies of logic have for us (Wittgenstein*). The so-called theory of types expresses just this, that certain verbal forms are not propositions. A clearer understanding of the role of 'proposition' in language is a primary requirement of mathematical philosophy. What then does the formalist axiom amount to? Clearly to this: that we shall *call* certain verbal forms 'propositions' even though they do not satisfy one of the characteristic properties of propositions (and this without rejecting that characteristic). As is to be expected, this leads to considerable puzzlement as we shall presently show. First, however, we shall try to see what it is that leads the formalist to take this step. If the decimal d had a finite number of digits, *given in extension*, then we could say that either one of the digits, at least, is unity or they are all zero. But in our case d is not given in extension; and to say that d has an *infinite extension* says only again that it is *not* given in extension. But the formalist falls into the common fallacy of thinking that although d has no physical infinite extension yet 'logically we may treat it as though it had'. Think of the familiar arguments about developing each successive digit in half the time of its predecessor; the renowned example of Achilles and the Tortoise is rarely properly understood by mathematicians. That the mathematical sum of an infinite arithmetical series is finite, in no way disproves Zeno's argument; we have here an excellent example of a word-series with quite different proposition-roles in different calculi. We feel that the 'question', 'Does Achilles catch the tortoise?' can have *only one* answer; whereas we have neither one question nor one answer. That $1 + \frac{1}{2} + \frac{1}{4} + \frac{1}{8} + \ldots$ has a limit 2 is a mathematical convention which we can apply or not, as we please. (Nor, of course, do the results of an 'actual' race answer more than *one* possible interpretation of the 'question'.) A similar mistake is at the root of the nonsense which physicists talk about

* Wittgenstein's name in brackets here signifies that I owe the idea to him. For the form of expression, and its context, since I am not quoting from any written record, I alone must accept responsibility.

'length'. The relativitist, they tell us, has *discovered* that what we think of as a ruler of fixed length, has in fact different lengths according to its position, a difference which, of course, no *measurement* could reveal. Whereas, what should be said is that 'length' in the relativity theory has a different usage from that in our everyday-language. To dispute about which is the *real* length is as idle as to dispute whether the king's-move in chess, or the king's-move in draughts, is the *real* king's-move. (William James perceived this clearly when he discussed the problem of the squirrel going round the tree.) The formalist's error is a false analogy between the finite and the infinite case (Wittgenstein). Another important factor is that the finitist's own mistake leads the formalist to believe that without this axiom 'the greater part of classical mathematics must be rejected' (and this is not the case).

While it is true that a formal calculus frequently assists in detecting errors which are unnoticed in ordinary language, each formal calculus carries with it new sources of confusion. Brouwer's analysis of the term 'infinite' has nothing to do with meaning, in the sense in which the formalist talks of meaning. The separation of sentential forms into sentences and non-sentences is part of the formal calculus, and the primitive sentences about *sentences* are defining characteristics of sentences; in particular the *t.n.d.* is such a defining characteristic, and according to this characteristic the sentential form, $(n)P(n)$, where 'P(n)' is '$2n$ is a sum of 2 primes', is *no sentence*. Just as, before Lagrange *invented* the fractional calculus, the word series '$f(x)$ is the $3\frac{1}{2}$th derivative of x^2' was no sentence.

The finitist fails to give a satisfactory account of classical mathematics, for he *openly* rejects it and builds up another system. The formalist fails and also builds a *different* system but cannot see that his system *is* different. And quite apart from the fact that formalism is *not* classical mathematics (or a foundation of classical mathematics) it is not in itself a satisfactory system. Not simply because of the confusion in the treatment of propositions, for even a contradictory system may have its interest. (This Carnap fails to see, for though he rightly says that the transformation rules of a calculus may be chosen quite arbitrarily, he nevertheless seems to believe that a calculus must be non-contradictory.[3] Consider the following very simple game.

Write down in a row as many numbers as you please, in any order. By repeated interchange of any 2 numbers try to bring the numbers into their natural order from left to right, or vice-versa, but after each interchange add to the end of the row the smallest integer not yet written down. If a given row is brought to order from left to right, the row will be said to be provable, if brought to order from right to left, refutable. It is clear that a row may be both provable and refutable, yet the game can be played none the less.)

The finitist considered that what he objected to in a reductio ad absurdum proof in mathematics, was an appeal to the *t.n.d* which, being unjustifiable, invalidated such a form of proof. One can perhaps best express his unwillingness to accept this proof in this rather paradoxical form: The reductio ad absurdum argument assumes that a word-series *is* a sentence in order to prove that it is one. This expression is, however, by no means the whole truth. Finitist and formalist alike make the mistake of supposing that there is only *one* form of reductio ad absurdum proof in mathematics, whereas in fact it has many forms. E.g. a proof written backwards, as in the proof of such simple propositions as $(a+b)^2 \geqslant 4ab$; for if $(a+b)^2 < 4ab$, then $a^2 + b^2 < 2ab$, and so $(a-b)^2 < 0$, which is false. Here we require only to reverse the steps (changing the inequality sign) and we obtain a 'direct' proof; and clearly there is no special merit in making the change. Or again a 'reductio ad absurdum' may be a 'direct proof' of a theorem quite other than the one enunciated, as when, to prove that if (a_n) is positive and strictly decreasing, and Σa_n is convergent, then $na_n \to 0$, we *in fact* prove that if $na_n \nrightarrow 0$, then Σa_n is divergent (and a quite different piece of mathematics proves the stated theorem). As Wittgenstein has often said, if you want to see what is proved, look at the *proof* (as opposed to the enunciation of the theorem).

Classical mathematics is a very complicated collection of systems, the various systems having terms in common, but the usages of the terms being frequently vastly different in the different systems. In particular 'number', 'proof', 'rational', 'real', 'equation' are used in a multitude of different ways. This would not be a source of confusion were it not for the temptation, that is strongly felt by mathematicians, to think that there is only *one* correct usage of a mathematical term, only *one* criterion of validity. The very name Mathematical Analysis points out a confusion (cf. Spengler, *The Decline of the West*, Vol. I, p. 81), what Wittgenstein calls 'the bad and widespread fallacy that mathematics is only a kind of physics, the investigation of such objects as numbers'. It is this confusion which leads to errors about 'completeness' or 'incompleteness' in mathematics (e.g. that the real numbers fill up the 'gaps' between the rationals), to speaking of the infinite as 'somehow given all at once' (for how could one *analyse* that which is not before one?). When Hilbert constructed his 'Grundlagen der Geometrie' he did not thereby supply demonstrations of the theorems of Euclidean Geometry which were previously lacking—he constructed a new geometry; and if one calls Hilbert's geometry a derivative of the Euclidean, one must remember that it is by no means the only possible derivative. One might say that Euclid's use of the word 'proof' has gone out of fashion. When Lindemann proved that π was irrational, he did not thereby solve the Greek pseudo-problem of 'squaring the circle'. If one chooses to describe the piece of mathematics which Lindemann invented by saying that it proves you cannot square the circle, you have then given a use to this sentential form which it did not have before. And if a Hilbert of the days before Lindemann had said that 'the circle is squared' must be true or false, he would in no way have been vindicated by Lindemann's work, because in the language to which it then belonged, the word-series 'the circle is squared' was no sentence—and Lindemann invented a new language in which this word-series was given a role. Through how many changes has the usage of the term 'circle' passed since the time of Archimedes! The confusion between two systems of mathematics, one in which we prove 'that every polynomial has a root' and the other in which we 'solve equations', leads to such questions as 'How can we say that an equation has a root when we have no way of finding it?' We have confused two different usages of the sentence 'an equation has a root', i.e. we have confused *two different sentences*, and when we realise this we are no longer surprised at the contrast. That a particular word-series can have different sentence-roles in different systems of mathematics is readily seen by considering e.g. 'two lines in a plane meet in a point'. In projective geometry this is an axiom (more strictly, I should say 'is the sign of an axiom'); in Euclidean geometry it is a false proposition; in Generalised Cartesian Geometry it is a true proposition (and in two-dimensional geometry it is no sentence).

The failure to understand correctly the role of 'proposition' in classical mathematics has led the formalist into a strange separation of true and demonstrable propositions. In classical mathematics a sentence (apart from primitive sentences) is said to be true only when it is demonstrated, whereas the formalist system admits of (non-axiomatic) sentences that are true, but not demonstrable.[4] From the formalist standpoint Gödel demonstrates that it is not possible to prove, in the calculus, that the calculus is non-contradictory. One might then well consider that this should oblige the formalist also to abandon the *reductio ad absurdum* argument; for a proposition established by such an argument cannot then be shown to be non-contradictory, i.e. if $(p \cdot \sim q \cdot \supset \sim p) \supset q$, then we may be unable to refute $p \supset \sim q$. E.g. if we prove that a bounded positive monotonic increasing sequence is convergent, only by showing that a divergent positive monotonic increasing sequence is unbounded, we have not at the same time shown that it is impossible for a bounded, positive, monotonic increasing sequence to be divergent; and this, on Gödel's theorem, we may be unable to do.

It cannot be too strongly stressed that the formalism of Hilbert or Carnap fails to give an account of mathematics, because the language of the latter is infinitely richer and more diversified than that of the formal system. The formalist accepts only *one* form of proof of the so-called universal proposition $(n)p(n)$, viz., $p(o) \cdot p(n+1) \supset p(n)$, whereas in mathematics we do not find this uniqueness. Consider, for example, the determination of the sum to n terms of the geometrical progression whose first term is unity and common ratio r. Denoting the sum to an unspecified number of terms, n, by s_n, we shall consider two of the many proofs to be found in mathematical works that $s_n = 1 - r^n / 1 - r$.

(1) For any assigned value of n, $s_n = 1 + r + r^2 + \ldots + r^{n-1}$, and so $rs_n = r + r^2 + \ldots + r^{n-1} + r^n$. By subtracting, we find $s_n(1 - r) = 1 - r^n$, and so $s_n = (1 - r^n)/(1 - r)$.

(2) Let '$p(n)$' be '$s_n = (1 - r^n)/(1 - r)$'. If for some n, $s_n = (1 - r^n)/(1 - r)$, then

$$s_{n+1} = s_n + r^n = (1 - r^n)/(1 - r) + r^n = (1 - r^{n+1})/(1 - r);$$

but s_1 and $(1 - r)/(1 - r)$ are both equal to unity, thus we have: $p(1)$ and $p(n) \supset p(n + 1)$, which demonstrates '$(n)p(n)$'. The formalist would accept only the second as a *proof* of the proposition $(n)p(n)$, and would say that, whereas the first proof shows that $s_n = (1 - r^n)/(1 - r)$ *for any assigned* n, the second alone demonstrates the universal proposition. Let us examine the second proof. We have demonstrated the propositions '$p(1)$' and '$p(n) \supset p(n + 1)$'. We have therefore set up a proof schema to establish $p(n)$ for any assigned n; for if we assign a number m, what our proof shows is that we can pass from $p(1)$ to $p(2)$ and then to $p(3)$ and so up to the assigned $p(m)$; i.e. our proof proves $p(n)$ for *any assigned* n. And if we call this the proof of '$(n)p(n)$' then we thereby lay down the equivalence of '$p(n)$ for all n' and '$p(n)$ for any assigned n'. This is a point in which the formalist has been misled by his calculus. In fact Gödel[5] has constructed in the formalist calculus a proposition $p(n)$ which can be demonstrated for any assigned value of n, but not for all values of n (though it is said to be true for all values of n). Our analysis of the universal proposition does not make this a contradiction, but shows again the extent of the divergence of formalism from classical mathematics.

Weyl has suggested that Brouwer's analysis of universal and existential propositions could be interpreted in a way that did not involve the rejection of the '*t.n.d.*'. His suggestion is that we should not equate '$\sim(n)p(n)$' with '$\exists n \,.\, p(n)$'. Like Brouwer's own interpretation, Weyl's misses the fundamental point that the crux of the question is a distinction between sentential forms and sentences. Were it merely a question of whether our calculus equates '$\sim(n)p(n)$' with '$\exists n \,.\, p(n)$' or not, *convenience* would be our only criterion of choice (as Carnap rightly here, insists). The problems we are considering are met with not only in mathematical analysis, but also in our everyday language, and they arise in cases which have no reference to an 'infinite set'. Let us construct an example. A man picks up a box containing coloured marbles and says: 'If there is at least one blue marble in this box, I pay £2 to a hospital, otherwise I pay £4.' In fact he pays £4 (and observes the rules). Can we say that there was no blue marble in the box? Clearly not, for if there was a marble of which he was unable to decide whether it was blue or mauve (say), he would be obliged to pay £4—and this *without* making the decision that the marble was not blue. Paying £4 does not constitute a decision on this question. To say that the marble must either be blue or not blue, says only if it says anything, that we could if we wish decide the question, (by an arbitrary decision *ad hoc*, i.e. by a definition), but this by no means changes the argument, which is: paying £4 does *not* constitute a decision; and if we choose to call paying £4 a decision that there was no blue marble, when in fact there was no such decision, then we have simply given the words 'to make a decision' the usage that the words 'not to make a decision' have in our everyday language. Moreover if, subsequently, we do make a decision, the payment of the £4 is irrelevant to that decision. Instead of a box of marbles we could consider a box of springs, and express the example in the form 'If there is a spring in the box over 3 inches in length, I pay £2, otherwise I pay £4.' Again no conclusion can be drawn from a payment of £4, for it may happen that all but one of the springs is under 3 inches, and this one oscillates so that it cannot be measured. The sentential series 'the length of the spring is 3 inches' is not a sentence in every context. For there may be no *accepted* way of measuring the spring and in such a case the sentential form would be of a kind analogous to 'The length of the colour blue is 3 inches'.

The following example seems to me to throw a clear light on the formalist calculus. The ordered set of integers (x, y, z, r) can be arranged in a sequence, (a_n) say. Let '$f(n)$' assert that, for the set of non-zero values (x, y, z, r) represented by the term a_n,

$$x^{r+2} + y^{r+2} \neq z^{r+2}.$$

Then we shall show that *classically*, $\exists k$ such that

$$f(1) \cdot f(2) \ldots f(k) \supset (n)f(n),$$

i.e. the 'proof' of Fermat's theorem, depends only upon testing a 'finite' number of cases.

We construct a sequence (S_n) as follows:

$S_n = .\alpha_1{}^n \alpha_2{}^n \ldots \alpha_n{}^n$, where $\alpha_r{}^n = 1$ for $r = 2, 3, \ldots n$, and, for all n, $\alpha_1{}^n = 0$ if $f(r)$ is true for all r, $1 \leqslant r \leqslant n$, but $\alpha_1{}^n = 1$ if $f(r)$ is false for any r, $1 \leqslant r \leqslant n$.

Thus the sequence (S_n) is strictly increasing and bounded above by e.g. .3. Now there is a theorem of the formalist calculus (based on a reductio ad absurdum) which says that a strictly increasing bounded sequence is convergent. Hence $\exists n$ such that $S_{n+p} - S_n < 1/10$ for all positive p. If $f(r)$ is true for all r up to and including this value n, then $\alpha_1{}^n = 0$, and so since $S_{n+p} - S_n < 1/10$, $\alpha_1{}^{n+p} = 0$ for all positive p, i.e. $f(n+p)$ is true for all p. We appear to have demonstrated that $\exists n$ such that $f(1) . f(2) \ldots f(n) \supset (n)f(n)$. Now it is obvious that no mathematician would consider that *anything* has been proved. We know no more about the possibility of proving '$(n)f(n)$' than we did before. The sentence 'there is a value of n for which $f(1) . f(2) \ldots f(n) \supset (n)f(n)$, but we don't know which it is', is precisely equivalent to: no matter how great a value of n for which we prove $f(n)$, we can say *nothing* about the truth or falsity of the 'proposition' $(n)f(n)$. And just this is the outcome of formalist metaphysics.

Let us consider in greater detail the sequence (S_n) we defined above. We showed that (S_n) was monotonic increasing and bounded above, yet no matter how great a value of n we consider, if $f(r)$ is true for all r up to that value of n, then it is *not* determinate whether $S_{n+p} - S_n < 1/10$ for all p, or whether there *is* a value of p for which this inequality is false. Thus, in spite of the formalists 'reductio ad absurdum proof' of the contrary, the sequence (S_n) is not demonstrably convergent. Now the formalist would oppose this contention in his usual way and say 'the theorem $(n)f(n)$ must either be true or false, and in *either* case the sequence (S_n) is convergent'. Here we see clearly the speciousness of this argument. For in the first place the argument admits that the 'reductio ad absurdum' is insufficient (i.e. irrelevant), and then bases the proof of convergence upon the *solvability* of another problem (Fermat's theorem in fact); in the second place, this self-same 'solvability' is an appeal to that very prin-

G

ciple (of sentence formation) upon which the (admittedly insufficient) 'reductio ad absurdum' was based.

This takes us to another great source of the formalist's confusions: his mistaken conception of the 'unsolved problems' of mathematics. That a certain word-series has the sentential form and contains only the symbols of a certain calculus, does not make that word-series a *sentence* in that calculus. To illustrate this we might consider certain well-known theorems in mathematics which, in the past, had the status of unsolved problems. First, however, let us take a fictitious case. Suppose a Newtonian analyst of the middle eighteenth century had proposed the 'problem', 'What is the 5/6th derivative of x^2?'. This might have been regarded by mathematicians as a *problem* which was *solved* by Lagrange in the latter part of that century. In this simple case such a misconception is not, in fact, likely to occur; few mathematicians (to-day) would fail to see that 'What is the 5/6th derivative of x^2?' is no question in the Newtonian Calculus, that, in fact, a *new language* is required before this question can be asked, (even though all the symbols in the word-series are familiar), and that Lagrange's 'fractional calculus' was a new language in precisely this sense. It is however far harder to appreciate that the so-called unsolved problems of our own time are not capable of being formulated in our language, and that, if they express anything, they express the *wish for a new language*. We are so familiar with 'Fermat's last theorem', with its romantic history, and with the fascination it has for mathematicians and laymen alike, that we feel irresistibly drawn to maintain that here we *have* a problem in the most ordinary sense of the term. For, on the one hand, Fermat himself claimed to have 'proved it', and, on the other, a whole literature of 'partial solutions' has grown up around it. Yet consider what *in fact* has been proved. *In the mathematical system in which the problem was supposed to be framed, practically nothing was achieved.* A totally new mathematic, the so-called 'theory of ideals' was *invented*, in which a problem, *of which Fermat had no conception*, was dealt with, and this great work is called paradoxically 'a contribution towards a solution of Fermat's problem'. In mathematics it is the solution which creates the problem. And I would advise the mathematician who fails to comprehend this to read Abel's

'Demonstration of the non-solvability by radicals of the general equation of higher degree than the fourth'.

The antinomies of logic are, basically, the outcome of the same confusions which Brouwer exposed in his analysis of the finite, and their resolution is both extremely simple and extremely difficult; for though it is easy to point out that here a definition is lacking and there a definition changed, that alone will not break the spell which the antinomy casts upon us. One must painstakingly trace the inter-connections and the relations of the antinomies, both to one another, and to other questions in mathematical logic. It by no means lessens the interest and value, from our point of view, of the paradox of the barber that it is a 'simple exercise in logic', as Mr. Grelling has pointed out,[7] to show that this paradox is the outcome of a faulty definition. For the 'barber' paradox is a derivative of the more famous Burali-Forti paradox. If one makes the class of men shaved by the barber *determinate*, i.e. if one decides *ab initio* whether or not the barber himself is a member of this class, then the apparent paradox either vanishes (when the barber is not included) or is seen to be the outcome of a contradictory definition. But it is, in a sense, the indeterminateness of the class 'all the men shaved by the barber' that gives the paradox its puzzling power. The proposition 'the barber shaves *all* those who do not shave themselves' seems straightforward when we first hear it, because we do not think of the barber as a member of the class of men shaved by him; then when we unwittingly change this class so as to include the barber, the paradox results. One could express this by saying that the word *all* sets a trap for us, and by changing its significance half way through the demonstration of the paradox, we are caught. In this case the 'all' wavers between one finite value and another; in the case of the analogous paradox for infinite classes—the Burali-Forti—the source of the confusion is again the 'all'. Here the 'all' is an extremely vague term conceived in some negative sense of 'without exception'. And Grelling's comment on the 'barber paradox' applies with equal force in this case. It is upon the proposer of this paradox to *define* the class of *all* ordinals. If he includes the ordinal of the class in his definition of the class, then his definition is contradictory—and if he doesn't, then of course no paradox results.

There is no need here for a theory of types, but of close scrutiny of the role played in mathematics by the universal 'all'. This same 'paradox' might well have occurred at an earlier stage of mathematical history. Let us suppose that in the following context by 'class' we mean 'finite class'. It is clear that from every class of positive integers we can form a new integer greater than every term of the class (e.g. the sum of the terms of the class). Consider, then, the class of *all* integers: this determines an integer greater than all integers—contradiction! We should not be deceived by this example, yet it contains all the elements of the 'series of all ordinals' paradox. What would be observed at once is that we cannot (in this language) talk of the class of *all* integers. Analogous reasoning throws some light on Cantor's proof 'that the class of all real numbers is non-enumerable'. From a given sequence of real numbers, he constructs another sequence of real numbers. Thus, we can determine as many sequences of real numbers as we please. This much the proof accomplishes; but to prove non-enumerability, we introduce 'the sequence of *all* decimals', apply the proof to this sequence and obtain a contradiction. This might well have been regarded, not as a reductio ad absurdum proof, but as a paradox. We have again been trapped by an 'all'. Notice that the *proof* says nothing about 'all decimals'. How, then, can we apply this proof to the 'sequence of *all* decimals' until this 'sequence' has been determined? There are many difficulties bound up with this question. In mathematics we meet with the 'all' of a finite class, the 'all' of an enumerable sequence, the 'all' of real numbers, of transfinite ordinals, etc. That the same word-sign has such a great variety of utterly different usages will inevitably lead to misunderstandings owing to our temptation to look for a unique usage, the real or true usage; what I think I. A. Richards not inaptly calls the 'one true meaning' superstition.

Quite apart from the Burali-Forti paradox, the current interpretation of the theory of transfinite ordinals is fundamentally metaphysical. We find in expressions of this theory such sentences as 'the ordinal ω *comes after all* the finite ordinals' (represented diagrammatically by a few finite ordinals, some dots, and then ω: observe that I am not criticising the notation but the interpretation of it. As one mode of representing a certain serial law this notation will perhaps do as well as any other—except that it has clearly led to such significant misconceptions in interpretation.) The root of the confusion about transfinite ordinals lies in the mathematician's confusion of intervals with classes of real numbers. Brouwer appreciated the difficulties about the definition of the class of all decimals, but did not see down to the source of the difficulty. Though it is very probable that he did perceive the relation of the 'class of all decimals', regarded as the 'class of all subclasses of a sequence', to his 'free-choice sequences'.

The extent of the bewilderment and confusion which results from a vague and uncritical usage of the terms 'class' and 'all' is well exhibited in Mr Russell's paper, 'The Limits of Empiricism'.[8] I choose this paper, not because it is unique in this respect (far from it), but simply because it is representative of a wide school of thought, and because Mr Russell's clarity of style renders it particularly useful for my purpose. We find there, for example, the sentence[9] 'Outside mathematics, we do not know with any certainty whether classes are finite or infinite, except in a few cases'. Now 'finite class' and 'infinite class' are terms which have certain usages in mathematics, which coincide only to a small degree with the usages in our everyday language. Just as 'infinitely happy' or 'infinitely rich' have but a distant relationship to the 'infinity' of mathematics. Russell speaks as though the relation of 'infinite class' to 'finite class' is something like the relation of blue to mauve, something about which we might easily be mistaken, rather than (say) the relation of 'end' to 'no-end', which would be nearer to the truth. When he proceeds to discuss the proposition 'all men are mortal' he shows, in fact, how vague and indeterminate is the 'all' of this proposition, and yet believes himself to be showing how clear cut and well defined the 'class of all men' is. In considering the question of the verifiability of this proposition, he comes to the conclusion that we could neither verify nor disprove it; thinking that there must be some *one* verification or disproof, instead of a multitude of different verifications, exhibiting the many different usages of this proposition. Mr Russell would not consider it impossible to enumerate some of the contexts in which he would use this proposition; that is all the finitist (or anyone else) need ask him for. One might say with William James, that if it *is* impossible to verify or disprove this proposition, (or, as it should be expressed, if there is no verification or disproof which Mr Russell would accept) then in no way would the world be different, whether the proposition 'all men are mortal' were true or not. And this would, of course, be the case if 'all men are mortal' was treated, not as a proposition, but as a *defining characteristic* of 'men'.

An analysis of Grelling's paradox[10] helps to clarify some of the confusions met with in the theory of types. The paradox may be stated in two rather different ways:

(a) A property p is heterological if the word 'p' has not the property p; thus 'heterological' is both heterological and not heterological.

(b) A word-property p, etc.

In the second case (and this is the form in which Grelling himself stated the paradox—the first form seems to be due to Carnap), the paradox is not very convincing. For 'heterological' is defined as a property of properties, and not as a word-property; or, looking at the question from another point of view, 'word-property' is a vague term in general usage, and there are many properties of which we should be in doubt whether to call them word-properties or not. We may, e.g. ask whether 'word-property' applies to the written, or to the spoken word, or to both. In particular, we may ask whether 'heterological' is a word-property; the answer to such a question must be a definition of 'word-property', and if this definition makes heterological a word-property, it is a contradictory definition, etc. The first form of the paradox contains rather deeper problems. The paradox in this form really makes 'property' vacillate between a quite general 'property', and a more specific 'word-property'. To say 'a property p is heterological if the word "p" has not the property p', leads us to think of 'p' as a word-property. To take 'heterological' as a value of 'p' shifts the meaning from 'word-property' to a wider 'property'. (I mean by this that in answer to the question 'Why can you take heterological as one of the properties in the definition?' the answer would be 'Well, heterological *is* a property'.) We are back to the crux of the finitist question, for we are here tempted to say that, surely, 'heterological' *must* be either heterological or not heterological. But should we be so willing to say that 'childless' must be either childless or not? (observe that the theory of types is silent on this point). The source of the difficulty is this: if there are a few examples (or even one) of a heterological adjective, then we suppose that '*x* is heterological' must be a *sentence* for every adjective '*x*'—in particular for '*x*' = 'heterological'—, whereas, in fact, the range of values of *x* is *only* the range that usage gives it. This is not to say that we cannot extend the range at will; we can, but the extension will change the definition of heterological, and we must not then be surprised if, after some change, we find that we have framed a contradictory definition. The very generality which the formalist regards as his main objective, is his greatest source of error. For example, in the Introduction to the *Logical Syntax of Language* (p. 2), Carnap writes '. . . given an appropriate rule, it can be proved that the word-series "Pirots karulize elatically" is a sentence, provided only that "Pirots" is known to be a substantive (in the plural), "karulize" a verb (in the third person plural), and "elatically" an adverb'. If 'Pirots karulize elatically' is a *sentence* for these reasons alone, it is a sentence with infinite freedom in our language, for nothing which we have been told about this word-series determines its position in relation to other word-series; it has been assigned no *role* in our language and we can make no use of it. 'I saw' and 'electron' are familiar words in our language, but we can do nothing with the sentential form 'I saw an electron' until we know just what is the experiment in connection with which this sentential form is used. Only when this experiment is known can we relate 'I saw an electron' to other propositions of the form 'I saw *x*', and only by its connection with other propositions is it given a sentence-role in our language. In mathematics, it is 'proof' which links a sentential form to the body of mathematical propositions. But we must not make the mistake of thinking that there is some unique form of proof. There is no standard of proof in mathematics, no single common factor, no quintessence of proof, only a great diversity of proofs; and the 'proofs' of one generation are but 'fallacies' to the next. The formalist endeavours to cloak the many forms of proof under a single rigid garment; the real task of the mathematical philosopher is to take off this garment and to study the diversity of forms he finds beneath. It is not a *new foundation* of mathematics that is needed but a close examination of its skeletal structure and of its ornamental coverings. Mathematics is like a city of fine buildings, filled with precious gems, but buried deep in the mud and sand of a desert. The task of digging up these treasures is a slow and arduous one; some progress has been made, but an account of the method and achievements of this work is outside the scope of this paper.

REFERENCES

1. Ambrose, Alice. 'Finitism in Mathematics', *Mind*, XLIV, No. 174.
2. Gödel, K. 'Über formal unentscheidbare Sätze der Principia Mathematica', *Monatshefte für Math. und Phys.*, 1931.
3. Carnap, R. *Logical Syntax of Language*, p. xv.
4. ——*Ibid.*, p. 133.
5. Gödel, K. 'Ünentscheidbare Sätze', *Monatshefte für Math. and Phys.*, 1931; Herbrand, J. 'Sur La Non-Contradiction', *Journal für Mathematik*, 1931.
6. Bouligand, G. 'Leçons sur La Théorie Generale des Groupes'.
7. Grelling, K. 'The Logical Paradoxes', *Mind*, XLIV, No. 174.
8. Russell, B. A. W. 'The Limits of Empiricism', *Proceeds of the Aristotelian Society*, 1935-36.
9. —— *Ibid.*, p. 144.
10. Grelling, K. 'The Logical Paradoxes', *Mind*, XLV, No. 180.

VII

THE NATURE OF MATHEMATICS

THE problem of the nature of the entities of mathematics continues to be, as it has been for the past hundred years, one of the central questions in foundation researches. Whether it is considered in its full generality or in the limited aspect of the existence problem, the question leads immediately to the heart of the controversy between formalism and finitism, realism and platonism.

The traditional view of existence in mathematics was summed up by Poincaré, at the turn of the century, when he said that a mathematical entity *exists* if its non-existence is impossible; in formal terms $\daleth(\forall x)\daleth P(x) \rightarrow (\exists x)P(x)$. The first (in this century) to deny the universal validity of this thesis was L. E. J. Brouwer. Brouwer utilised current unsolved problems in number theory to show that the existence of a mathematical entity in Poincaré's sense may depend upon the discovery of the solution of some hitherto unsolved problem, and this discovery may never be made. The essence of Brouwer's argument may be expressed in the following form.

Let $P(n)$ affirm some property of n of which it is unknown whether every natural number has this property or not; e.g. $P(n)$ may be the property that $2n$ is a sum of two primes. We may obviously choose for $P(n)$ a property for which a definite decision procedure exists by which we can tell whether any individual n has this property or not; in fact the example cited fulfils this condition, and we may readily construct an unlimited number of such examples (not necessarily of any intrinsic interest). With $P(n)$ we associate a function $p(n)$ defined as follows:

$$(\forall r)\{r \leqslant n \rightarrow P(r)\} \rightarrow p(n) = 0$$
$$(\exists r)\{r \leqslant n \,\&\, \daleth P(r)\} \rightarrow p(n) = 1$$

so that $p(n)$ is always 0 or 1, and is non-decreasing. The limitation we imposed upon P ensures that the value of $p(n)$, for each assigned **n**, is as definite as the procedure by which the question whether **n** has the property P or not is decided. Since $p(n)$ is non-decreasing and bounded a familiar theorem of classical analysis affirms that $\lim_{n \to \infty} p(n)$ exists; this limit must be either 0 or 1, since $p(n)$ takes no other value. If

$$\lim_{n \to \infty} p(n) = 0 \text{ then } (\forall n)\{p(n) = 0\},$$

so that $(\forall n)P(n)$ holds, and if $\lim_{n \to \infty} p(n) = 1$ then $(\exists r)\urcorner P(r)$ holds. Thus a knowledge of the limit of the sequence $p(n)$ solves the problem whether all numbers have the property $P(n)$ or not, and conversely a solution of this problem supplies us with the limit of the sequence.

Now it is quite clear that the classical proof of the convergence of $p(n)$ does *not* tell us whether $P(n)$ is true for all n, or not. If the truth of $P(n)$ for all n is unknown before the convergence of $p(n)$ is proved, then it remains unknown after the convergence of $p(n)$ has been established. Brouwer sought to show by examples of this kind that certain proofs in mathematics, like the classical proof of the convergence of a bounded monotonic sequence, are unworthy of our confidence and should be discarded. Before we look more closely at the type of proof in question it is very informative to consider briefly a familiar objection which continues to be raised against Brouwer's critique. Fix our attention on some particular unsolved problem, it is said, and then Brouwer's dilemma disappears; for when the problem is eventually solved we shall know whether the limit of $p(n)$ is zero or unity. At present we can say that the limit *exists*, we just do not know *what* it is; as we might say of some equation, that it has root in a certain square but one cannot tell precisely where in the square it lies. We need more information to fix the limit, but we already know that it exists. There are two ways in which one may deal with this objection. First we may ask what *sense* it makes to say the problem will be solved some day. Consider the problem which Brouwer has often made use of, whether seven consecutive 7's occur in the decimal expansion of π. Does the expansion to 10^{10} places, say, exist before it has been calculated? As Wittgenstein has remarked, does a character exist in a play which has yet to be written? If, as we may suppose, the only method of attacking the problem is to proceed with the expansion of π then it is impossible to prove that a run of seven 7's does *not* occur; and the situation is not improved by attacking the problem with *all* the resources of current mathematics, for now we have merely replaced the question whether seven 7's occur in a certain expansion by the similar one whether or not a particular sentence occurs in the unending chain of inferences which we can draw from our present knowledge. Moreover, in virtue of results which have been obtained during the past 25 years, we can meet the objection in another way. In place of an unsolved problem let us now consider *any* property $P(n)$ in elementary number theory. Exactly as before we may associate with $P(n)$ a sequence $p(n)$ such that $\lim_{n \to \infty} p(n)$ exists classically and has the value 0 or 1 according as $P(n)$ is universally valid or not. Then, *if* the classical convergence proof actually provided us with the limit of $p(n)$ we should have a decision procedure for elementary number theory; that is to say if the proof of convergence provided a method of *calculating* the limit, this method would constitute a decision procedure for number theory. Since we know from Church's theorem that no decision method for number theory exists (assuming consistency) it follows that it is *impossible for the convergence proof to provide a method for calculating the limit.*

The better to see the significance of this result, let us concentrate upon a particular class of properties, namely primitive recursive properties $P(n)$, for which the associated function $p(n)$ is primitive recursive. The associated functions, being primitive recursive are enumerable by a doubly recursive function $I_n(m)$ say, where $I_n(m)$ is the mth function in the enumeration. Then the result of the classical convergence proof is that for each m

$$\lim_{n \to \infty} I_n(m)$$

exists and has the value 0 or the value 1. Now if the function $f(m) = \lim_{n \to \infty} I_n(m)$ were a *computable* function (in Turing's sense, for instance) then $f(m)$ would be a decision function for the class of primitive recursive predicates (of one variable) but it is known* that this class does not admit a computable decision

* Reference [3], pp. 417-418.

function. Thus the *function* $\lim_{n \to \infty} I_n(m)$ given by the convergence
proof is *not* computable, that is to say this 'function' admits no
computation procedure. The question is now seen in its proper
setting. What is involved is not a matter of *existence* but of *word
usage*. Shall we call $\lim_{n \to \infty} I_n(m)$ a *function* or not? Classical analysis
accepts $\lim_{n \to \infty} I_n(m)$ as a 'function', but Brouwerian mathematics
does not. *It is not whether or not a certain entity exists which is in
dispute but how the term function should be used.* Brouwer's own
presentation of the issue obscures this fundamental fact, for, as
we shall have further occasion to observe, a failure to clarify the
function concept vitiates much of intuitionist analysis.

Classical analysis is not rendered demonstrably contradictory
by its current use of the term function. It may be held to be
absurd to call a term for which no computation procedure is
logically possible a *function*, but it is not contradictory. Whether
one continues to use a system of mathematics which uses the
term function in this absurd way is of course a matter of
individual choice.

It is often said that the intolerable complexity of revisionist
systems of analysis far outweighs any advantage they may have
on purely rationalist grounds, but this objection appears to be
based as much on a want of knowledge of some branches of
modern mathematics as on an unfamiliarity with the new
systems. A more serious objection, at first sight, is that the
classical usage of function has led to no difficulty in application;
if there were some intrinsic absurdity in this usage would it not
have shown itself in some Engineering or Laboratory experiment? For instance, the laws of arithmetic are verified countless
times in commerce and science every day, and at a higher level,
the conclusions of classical analysis have found innumerable
applications without any adverse consequence. This argument
from application is an important one on which we shall have
more to say later, but for the present we note that no experiment
is crucial in deciding between the classical and the finitist
systems. For all experimental purposes it suffices to take any real
number to, say, 10 places of decimals; to replace any integral by
a finite sum; any function by a rational approximation; and so

on. The argument from application in fact is on the side of such systems (as recursive analysis) which concentrate upon the applicable parts of mathematics and operate in the rational field.

The classical convergence proof to which we shall now return consists essentially in the following steps:

We have to prove the formula

$$(\exists k)(\forall n)\{p(k+n)=p(k)\}. \tag{F}$$

From its contrary

$$(\forall k)(\exists n)\{p(k+n)\neq p(k)\},$$

noting that by definition $p(k+n)\geqslant p(k)$, we deduce

$$(\exists n)\{p(k+n)>p(k)\};$$

hence if we denote by $\theta(k)$ the *least* n for which this inequality holds we have both

$$p(\theta(0))>p(0)$$

and

$$p(\theta(0))+\theta(\theta(0))>p(\theta(0))$$

whence $p(\theta(0))\geqslant 1$ and $p(\theta(0)+\theta(\theta(0)))\geqslant 2$, the last of which contradicts the defining condition $p(n)\leqslant 1$, completing the proof of (F). To break the proof Brouwer rejected one part of the equivalence

$$\rightarrow(\forall x)P(x)\longleftrightarrow(\exists x)\rightarrow P(x)$$

namely, the implication

$$\rightarrow(\forall x)P(x)\rightarrow(\exists x)\rightarrow P(x), \tag{E}$$

whilst retaining the converse implication

$$\rightarrow(\exists x)\rightarrow P(x)\rightarrow(\forall x)P(x). \tag{A}$$

The grounds for rejecting (E) are that a disproof of $(\forall x)P(x)$, for instance by deriving from it a contradiction, does not provide any means of finding a counter example to the universal statement $(\forall x)P(x)$. One of the difficulties of dealing with an existence proof in this way, is that while retaining the existence operator, we have provided no means of proving existence; or rather, we have left unanswered the question *what* constitutes a valid existence proof.

For instance, in the definition of a null sequence, in place of the classical definition that a_n is a null sequence if

$$(\forall k)(\exists n)(\forall p)\{|a_{n+p}|<1/(k+1)\}$$

holds, the intuitionist says that a_n is a null sequence if there is an effectively determined function $n(k)$ such that

$$n \geqslant n(k) \rightarrow |a_n| < 1/(k+1);$$

but as to what functions are effectively given the intuitionist remains silent. This is another example of the failure to analyse the function concept to which we have already referred.

Following suggestions of Herbrand and Gödel, attempts have been made to identify the notion of an effectively defined function with three equivalent concepts: Turing computable functions, λ-definable functions and quasi-recursive functions (known also as general recursive functions). A quasi-recursive function is one whose values may all be obtained by repeated substitution in a finite number of defining equations. No limit is placed upon the number of substitutions which may be needed, and therein lies one of the weaknesses in this attempted characterisation of effectively defined functions. What is meant by a finite number of substitutions in this context? For each value of n, we require that the value of $f(n)$ be determined by a finite number of substitutions. If we denote by $V(n)$ the number of substitutions needed to determine $f(n)$ for each value of n we see that a function lies concealed beneath the term 'a finite number'. The definition of quasi-recursiveness gives us no information about the function $V(n)$ and for lack of this information a quasi-recursive function is not effectively defined. Would anyone undertake to pay a computer by the hour to evaluate even $f(0)$, just on the assurance that it will take only a finite time to complete the calculation? As a London Park-keeper had to decide recently, how long may a dog's lead be before the dog ceases to be under control? The weakness of the attempted identification of quasi-recursiveness with effective definition is brought out even more clearly by considering a certain sufficient condition for a function to be quasi-recursive. To formulate this condition we start by defining the class of primitive recursive functions (to which we have already had occasion to refer).

A function $f(a, n)$ is said to be *defined by primitive recursion* from functions $\alpha(n)$, $\beta(a, n)$ if

$$f(a, 0) = \alpha(a), f(a, n+1) = \beta(a, n, f(a, n))$$

(in this schema none of the functions contains a concealed para-meter, but some or all of the parameters in α, β may be absent and then f may depend only upon n); a function (of any number of variables) is said to be primitive recursive if it is one of the initial functions $Z(n) = 0$, $I(n) = n$, $S(n) = n+1$ or is defined by substitution or recursion from primitive recursive functions.

It was shown by Kleene ([4], p. 279) that a sufficient con-dition for a function $f(x)$ to be quasi-recursive is that there should be some primitive recursive relation $R(x, y)$ for which $(\exists y)R(x, y)$ holds (in some formal system S, say) and $f(x) = (\mu y)R(x, y)$. Now the condition $(\exists y)R(x, y)$ may be estab-lished (in S) by showing that $(\forall y) \to R(x, y)$ leads to a contradic-tion and so we may have no means of finding the least y for which $R(x, y)$ holds for a given x, except by testing the predicates $R(x, 0)$, $R(x, 1)$, $R(x, 2)$, ... in turn *without limit*. To say that, if S is free from contradiction, we shall eventually reach the predi-cate $R(x, p)$ which holds for the given x is certainly not to provide an effective procedure for finding it. Again we may ask how much will it cost to find it, and the fact that the question makes no sense here—there is no possibility of estimating cost—shows the nature of the problem. It may be said, and Church in fact said this ([1], footnote 10), that the function $f(x)$ is defined as constructively as the existence condition $(\exists y)R(x, y)$, but this appears to make the identification of 'quasi-recursive' with 'effectively definable' a vicious circle. For $(\exists y)R(x, y)$ is said to be established constructively if there is an effectively defined function $\theta(x)$ such that $R(x, \theta(x))$; and if therefore we say that $(\mu y)R(x, y)$ is effectively defined if $(\exists y)R(x, y)$ is established con-structively we are saying only that $(\mu y)R(x, y)$ is effectively defined if it *is* effectively defined. The notion of quasi-recursiveness may provide an upper bound to the class of effectively defined functions but it is too general to characterise that class. In fact the attempt to identify the notion of a *con-structive existence proof* with the existence of a constructible entity is untenable, for it is known ([5]) that for any continuous recur-sive real function F such that $F(0) = -1$, $F(1) = +1$ there exists a recursive real number α such that $F(\alpha) = 0$, but the existence of this α cannot be proved constructively, for as Specker has shown there is a sequence $F_k(x)$ of continuous recursive real functions such that for *any* recursive sequence of recursive real numbers α_k, $F_k(\alpha_k) \neq 0$ for some k.

Can a real number $.a_1a_2a_3a_4. \ldots$ be defined by a succession of free choices of 0 or 1? This question is vital to the development of intuitionist analysis, for the intuitionist continuum is derived from the notion of free choice sequences. What is the difference between say, defining a real number $.a_1a_2a_3. \ldots$ by the recursive conditions $a_1 = 0$, and

$a_{n+1} = 0$ if $a_n = 0$ and $2n + 2$ is a sum of two primes,
$a_{n+1} = 1$ if $a_n = 1$ or $2n + 2$ is not a sum of two primes,

(let us call this number the Goldbach number), and defining $.a_1a_2a_3. \ldots$ by a succession of free choices. I am not now concerned with the question whether the Goldbach number is zero, or not zero, but with the difference *in kind* of these two definitions, a difference which does not appear to have been properly appreciated. At first one might say that the difference is between proceeding according to a rule and proceeding without a rule, but this is not satisfactory because the concept of rule is ill-defined, and one may well say that in the free choice case the rule is to write down 0 or 1 whichever is the first to come into one's mind. The important difference is that the Goldbach number is in a certain sense, determined in advance by the recursive law of definition, whereas the free choice number is *free*. In one case we have the recursive law *and* the digit by digit development of a decimal. It seems to me that only in the first case can we speak of a real number, the recursive law itself, whereas in the second case we have, not a real number, but a daily changing collection of terminating decimals. If I add a digit a day (at random) to *any* decimal, then I am producing only finite decimals however long I may live, and if I leave the task to my successors then they too will produce only finite decimals, however many generations may follow me. And if a random machine is built to produce the decimal, then this too will produce only terminating decimals which will have an end in time. But the recursive *function* is here now, and is independent of the passage of time. There may cease to be men, or machines, to *use* the function, but the function itself is *timeless*. These observations make sense of course only if we accept definition by recursion as itself constituting the whole function. In the classical sense a function is a class of pairs (x, y), y being the value of the function for the argument x, and with this definition of function it makes no sense to distinguish between a real number growing digit by digit, and a real number specified by a function. But the recursive definition makes no reference to the values; it is a rule for operating with the sign $f(n)$ itself. The importance of thus distinguishing between a function and its values was first clearly brought out in Gödel's construction of an undecidable sentence where we find, in effect, a function $f(n)$ such that each of the equations $f(0) = 0, f(1) = 0, f(2) = 0, \ldots$ is provable in some system Z, but $f(n) = 0$, with free variable n, is not provable.

The intuitionists say that given the structure

$$.0 \quad , \quad .1$$
$$.00 \quad , \quad .01 \quad , \quad .10 \quad , \quad .11$$
$$.000, \quad .001, \quad .010, \quad .011, \quad .100, \quad .101, \quad .110, \quad .111$$

. . . .

in which the nth row is the sequence of all n-figure binary decimals (with digits 0, 1) in increasing order of magnitude, the continuum of real numbers may be obtained by thinking of real numbers as threads running through the structure, passing from a term in one row to a term in the next which has the same initial segment. But it seems to me that this structure, with or without threads, determines only terminating binary decimals; if we *have* a function $f(n)$ (let us for the moment suppose that $f(n)$ is primitive recursive) which chooses the nth digit of the decimal, then certainly we have a real number, but in this case the 'spread' of the decimals is irrelevant, and lacking a function we have nothing but the 'spread' itself. For what does it mean to say that we *can* choose a digit at each stage? Who chooses? And of course if we answer that we can give a law, then we must say what we mean by a law, and there will again be no use for the 'spread'—the law of choice will be the real number. In fact, to define a real number we must define a function, and we are back to the question of what constitutes an effective definition of a function.

As we have seen, quasi-recursiveness is too wide for our

H

purpose, and, of course, primitive recursiveness is too narrow. From single recursion we may pass to recursions in 2 variables, in 3 variables and so on up to any number of variables, without losing the fundamental character of a recursive definition, viz., that the value for an assigned argument is determined by an assignable number of substitutions. Then we may pass to transfinite recursions of ordinal ω, ω^ω, $\omega^{\omega\omega}$ and so on. In our present state of knowledge ([2]) we may proceed as far as $\omega^{\omega\omega}$ and retain the essential features of the definition. The difficulty here lies in an elimination of a transfinite induction. The principle of transfinite induction assures us that a decreasing sequence of ordinals is *finite*, so that the values of a transfinite recursive function defined, say, by the equations

$$f(0) = 1$$
$$f(n+1) = a(n, f(\gamma_\Omega(n)))$$

where $\gamma_\Omega(n)$ is a predecessor of $n+1$ in some recursive ordering of the natural numbers of ordinal Ω, may be obtained by a *finite* number of substitutions. As we observed with quasi-recursions the information given by transfinite induction, that the number of substitutions is finite, is not precise enough, and some specification of the number of substitutions has been found for decreasing sequences of ordinals not greater than $\omega^{\omega\omega}$, but not yet for greater ordinals. The first ϵ-number ϵ_0 appears to constitute a natural frontier. For an ordinal $\gamma < \epsilon_0$ it seems probable that we shall be able to prove that the number of terms in a decreasing sequence of ordinals, starting from γ, is at most a value of a function defined by a recursion of ordinal less than γ. This would enable us to show that the number of substitutions needed to obtain the value, for any assigned argument, of a function of ordinal γ is given by a function of smaller ordinal. It is known for instance ([2]) that for a function of ordinal ω^ω, the number of substitutions needed is given by a function of ordinal less than ω.

One of the reasons why ϵ_0 appears to be a natural barrier here is that, as was shown by Gentzen, transfinite induction over ordinals less than ϵ_0 may be replaced by induction over the natural numbers, but an induction over ordinals up to and including ϵ_0 cannot be so replaced (and that this is the case is

indirectly known from Gentzen's proof of freedom from contradiction of number theory in conjunction with Gödel's proof of
the impossibility of establishing this freedom from contradiction
by means of the resources of number theory alone).

We remarked earlier that in his criticism of the classical pure
existence proof Brouwer rejected the implication

$$\rightarrow(\forall x)P(x) \rightarrow (\exists x)\rightarrow P(x)$$

but retained the converse

$$\rightarrow(\exists x)P(x) \rightarrow (\forall x)\rightarrow P(x)$$

on the grounds that the impossibility of a counter example
assures universality. It seems to me that this principle is also
questionable. How does it establish $(\forall x)P(x)$ to show that the
hypothesis that a counter example exists leads to a contradiction?
presumably we test in turn $P(o), P(1), P(2), \ldots$ without finding
a case of failure. We then prove that if we chanced upon a case
of failure we should have found a contradiction (in Arithmetic),
whence we conclude $(\forall x)P(x)$. For want of a proof of freedom
from contradiction the argument is little more than an expression of faith in consistency. Brouwer considered that *reductio
ad absurdum* is the natural way of proving a *negative* property, for
instance that $\sqrt{2}$ is not rational. But the classification of properties into positive and negative is surely quite arbitrary. Are not
rational numbers the exceptions amongst the real numbers, and
could we not regard a proof of the irrationality of $\sqrt{2}$ as an
Existence proof, a proof that amongst the irrationals there is one
equal to $\sqrt{2}$? In fact *direct* proof of the irrationality of $\sqrt{2}$ is to
hand. It suffices to observe that $2q^2 - p^2$, $q \neq 0$, $p \neq 0$, necessarily
has an odd factor, to show that $|2q^2 - p^2| \geq 1$ and so that
$|2 - p^2/q^2| \geq 1/q^2$ which not only shows that p^2/q^2 differs from 2,
but tells us by what amount at least they differ. Is not '$|2 - p^2/q^2|$
$\geq 1/q^2$' a *positive statement*? Of course if we accept the conventional indirect proof (which uses the same expression $2q^2 - p^2$), it
follows from $2q^2 - p^2 \neq 0$ that $|2q^2 - p^2| \geq 1$ and so the bound
$1/q^2$ for the least difference between 2 and p^2/q^2 is readily found,
but is it not strange to have to derive $|2q^2 - p^2| \geq 1$ from
$2q^2 - p^2 \neq 0$, when in fact the same piece of mathematical reasoning (considering the factors of $2q^2 - p^2$) leads directly to
$|2q^2 - p^2| \geq 1$. Another weakness in the classical *reductio ad*

absurdum proof is that it starts by assuming that $\sqrt{2}$ is either rational, or irrational, whereas it is impossible to provide a decision procedure by which we may decide, of any real number, whether it is rational or not. A more interesting distinction between direct and indirect proof may be drawn when we consider rather more complicated examples.

Let $s(n)$ be a primitive recursive function (integral or rational) which is primitive-recursively convergent so that there is a primitive recursive function $N(r)$ such that

$$n \geqslant N(r) \to |s(n) - s(N(r))| < 2^{-r}.$$

A classical proof of irrationality of $s(n)$ establishes the quasi-recursive irrationality of $s(n)$, that is to say, that there are quasi-recursive functions $i(p, q), j(p, q)$ such that

$$n \geqslant j(p, q) \to |s(n) - p/(q+1)| \geqslant 2^{-i(p, q)},$$

and this is all the information that the classical proof yields. It can be shown however that algebraic numbers, and e^x for rational x, and π, are all *primitive-recursively* irrational, the associated functions i and j being primitive recursive. Direct proof not only merits more confidence but yields more information. The same situation arises when we consider transcendence; the classical indirect proof suffices to establish the quasi-recursive transcendence of e and π, but to prove primitive-recursive transcendence direct proofs are needed, and have been found by a more detailed analysis of the method employed in the indirect proof.*

* Doubts concerning the validity of the use of quantifiers and of indirect proof led me, in 1938, to the construction of the equation calculus, a free variable system in which recursive arithmetic and recursive analysis may be codified without postulating any logical axioms. Finding no appreciation of this system in England or the United States, I sent an account of it to Professor Bernays just before the outbreak of war and our subsequent correspondence travelled to and fro across Europe, no doubt unnecessarily worrying the Censorship of three countries.

I am glad to have this opportunity to record Professor Bernays' immediate mastery of the system exemplified by the new proof of theorems on substitution and induction which came to me almost by return of post. The encouragement which he then offered me gave me the strength of purpose to continue working during the dark days of the war, and like so many mathematicians, in Europe and in the United States, I never cease to be

One of the ways of gaining insight into the nature of the entities of mathematics is to trace the evolution of the concept of a formal system. Let us consider for example Euclidean Geometry. The theorems of geometry we may suppose originated in the 'building mathematics' of the Egyptians. The use of a loop of cord of 12 units length, divided into three parts of lengths 3, 4, and 5 units, to construct a right angle is the source of Pythagoras' theorem, and the practice of laying out a hexagon using a single cord of constant length to describe a circle and mark off the vertices of the hexagon, is one of the child's introductions to circle geometry to this day. At this stage geometry is all 'geometrical drawing'. The circle, line or triangle are physical marks, not concepts. The construction is *copied*, not recreated, each time it is used. An analogous situation in arithmetic is found in the work of the amateur mathematician. For instance an amateur mathematician recently made the following discovery (not of course in this notation).

Let $S_n^k(x)$ be the sum obtained by replacing each of the first k terms in the series $1 + 3 + 5 + \ldots + (2n - 1)$ by the number x, then the number $8N - 1$ is composite if and only if there are numbers k, n with $k < n$, such that

$$\tfrac{1}{2}N = S_n^k(\tfrac{1}{2}) \text{ or } \tfrac{1}{2}N = S_n^k(-\tfrac{1}{2}).$$

The discovery had presumably been made by an extensive study of particular cases and the discoverer had little knowledge of algebra or proof processes. Merely because the discovery is concerned with numbers does not of course make it mathematical; the man who screws the house number on the door is concerned with numbers, but not with mathematics. What makes the discovery mathematical is clearly that it adds something to the sum of mathematical knowledge, but it must be remembered that what is added is just the *specific instances* of the result which are shown to hold; the fact that the general result is provable (by quite other methods of course) does not make the original discovery something more general than in fact it was. We are often too ready to see the discovery of the general in the

grateful to Professor Bernays for his generous advice and assistance given so freely from his encyclopaedic knowledge, his great skill and the breadth and depth of his understanding.

discovery of the particular, by a false analogy with the physical sciences. If after 20 experiments we find that sulphur melts at 115° centigrade, and maintain that figure in the face of all subsequent experiments, then we are making melting at 115° a *defining property* of sulphur. One who maintained the universality of a property which held only for the first 20 numbers (saying for instance that all calculations which led to opposed results were necessarily erroneous) would not be held to be making a contribution to mathematics. Of course there is a situation in mathematics more closely analogous to the physical one. The 'laws' of arithmetic like $2 + 3 = 5$ were no doubt originally statements of observed fact and now have the force of convention, but this we shall come to shortly.

The second stage in the development of a formal system is the discovery of *connections* between the observed facts. For instance that the square on the hypotenuse may be cut up and rearranged to form the squares on the other two sides of a right angle triangle, or that the area of a triangle is equal to half the product of its height and base, and so on. From these *connections* the concepts are beginning to emerge. In the next stage the notion of a connected chain of theorems is dominant. Results depend upon one another, the notion of proof is implicit and 'reasonable' assumptions are sought for as starting points; but the important thing is that a body of facts are being woven together, and the formation of these links is as much a mathematical activity as the original observation of the 'facts'. Then the nature of the connections themselves becomes the subject of inquiry and proof is formalized and incorporated into the body of results. At the final stage all preconceptions about the necessary character of the proof processes and the origins of the subject matter are deliberately cast away and the formal system is born. Mathematics is not any one part of this evolutionary process, for the whole process of development, at each stage, is a mathematical activity. The creation of the formal system, *and* its manipulation, are mathematics. And the formal system itself is not the end stage in mathematics, for the process of making similarities and differences, of finding out general features and connections, now starts all over again, but this time the formal systems themselves are the object of study, instead of circles and numbers.

In the evolution of a formal system, what develops is the *idea* of proof. The contrast between an intuitive and a formal proof is often misunderstood. It is sometimes said that we have a preconceived intuitive notion of proof which *eventually* finds expression in formal proof, as if the formal proof was somehow already present in our mind and seeks only for expression. This may or may not be partially true as a matter of history, but it is a misleading picture from the point of view of the analysis of concepts. What is *accepted* as proof itself evolves. It is no more true to suppose that the concept of proof is inborn in man than to suppose that traffic regulations are inborn. Just as traffic regulations have perforce grown more and more involved and extensive as traffic has grown denser, so the fecundity of mathematical invention has made a more definite and formal proof concept necessary. Another common misunderstanding takes the opposed viewpoint that *ultimately* proof is not formalisable. This view arises from the contrast between the operation of a formal system and the proof of general results *about* the system. The difference is simply that between proof in *one* system and proof in *another*. The fact that the second system—the meta-system—may not be formalised is an inessential difference. We are still operating a system with rules even if we have not taken the trouble to make the rules precise. The distinction between a formal system without meaning at one level and a meaningful metamathematics at the next seems to me quite untenable.

Fundamentally mathematics is a concept creating activity. The mathematician does not study the world as nature, he is not classifying his experiences, like a Zoologist. The triangles and circles of geometry are *created by mathematics*. Certainly it is true that triangular objects and circular objects are part of our everyday experience, and these objects may *serve* as the triangle and circle of geometry. But when they so serve it is not their physical properties that we study. We use them to draw attention to certain general features, to certain connections and to certain differences. We use them, not as physical objects, but as symbols. Precisely the same situation occurs in language when we use certain shapes—or sounds—as words.

If mathematics had no applications then it would indeed be 'merely a game'. Even so, not a meaningless game, as is sometimes said, for operating with symbols in a coherent structure is itself meaningful; language itself is nothing else than operating with symbols. But the application of mathematics is one part of mathematics and the meaning of its concepts reflects this application. Before arithmetic became a formal system, its theorems derivable from axioms by formalised rules of procedure, the sentences of arithmetic, as we have already observed, expressed the facts of experience. Two apples and three apples make five apples, and so on. In the course of time this led to the formalisation of the rule $2 + 3 = 5$, and 'two apples and three apples make five apples' became an application of the rule. The fact that drops of water may not behave in accordance with the rule, is not allowed to challenge the rule *qua rule*; the arithmetic of apples is not the arithmetic of water droplets. That $2 + 3 = 5$ is derivable from suitable definitions of addition and of the numbers 2, 3 and 5 is *another* mathematical discovery. Formal arithmetic grew out of the everyday arithmetic of apples and pears, by allowing the expression of a fact to fossilise into a convention —a process that is always misleadingly called abstraction, as if the rule were somehow contained in the fact and then drawn out.

Of all the entities of mathematics the oldest and the most important of course is the natural number; number is one of the chains which binds mathematics to the real world. Nevertheless, numbers are dispensable, and everything which can be said by means of number words can be said without them—the sacrifice of the number words involves language in well-nigh intolerable complications, but the words are dispensable. To say that a room has three windows is just to say that it has a window and a window and a window. And to say that it has two doors is to say that it has a door and a door. The number concept is created by treating 'a window and a window and a window' as an 'application instance' of the rule '$x + x + x = (1 + 1 + 1)x$' by means of which 'a window and a window and a window' is transformed into 'one and one and one window'. The final step of introducing the abbreviations 'one and one is two', 'two and one is three' is not essential to the generation of the number concept. It is this last step which is performed by counting, so that counting does not determine the number of a class but transforms that number from one notation to another. Counting is a process of translation not of discovery. The difference, between number signs and other words is that number signs are spelt by a rule. Number signs (we may say) are words 'spelt' with the letter 1, and this letter alone. This is all that we mean when we say that the class of numbers is *infinite*; that we impose no limit to the length of number words. Any word spelt only with 1's is a number sign, like 1111 or 111111 (we are not now thinking of a positional notation in a scale). All this is clear enough; the difficulty starts when we look for that of which the number-sign is a *sign*. By analogy with chairs and the word 'chair' we look for that which the word is supposed to denote. But to use Frege's pregnant distinction, words may have either sense, or denotation, or both. The denotation of 'chair' is chair, but this is not the sense of the word, and the physical object is not the concept. Frege was unable to explain the origin of the sense of a word, and it was Wittgenstein who first showed us that what gives a word 'sense' is the part which it plays in the language to which it belongs. The sense of the number words, the numerals, is to be found only in arithmetic itself. Frege and Bertrand Russell both sought a denotation for number words in classes of similar classes, but they found only a new concept. The intuitionists choose to found arithmetic on psychology, on our supposed fundamental intuition. The fallacy of basing mathematics on psychology has been brilliantly exposed in Frege's writing; and if Frege's words are not enough we may add the evidence of recent psychological experiments which appear to show the young child to be devoid of the number sense. It would seem then that the number concept is to be found only in the transformation rules of arithmetic, and the use of collections as number signs. The question 'what is number' must be replaced by the wider question 'what is arithmetic and what are its applications'.

Frege and Russell criticised the formalist treatment of number, the view that the number concept derives from the transformation rules of the number signs, by saying that this view resulted from a confusion of number with numeral. That $2 + 3 = 5$ is not an assertion about numerals, since clearly '$2 + 3$' is not the same sign as '5'. But Frege's criticism is based in part on a misunderstanding of the equality sign. In '$2 + 3 = 5$' the equality sign expresses the fact that the sign '2 + 3' is a *transform* of the sign '5'. The two sentences 'the sum of the numbers two and three is the number five' and 'the numeral "5" is a transform of the sign "2 + 3" ' have the same meaning, an instance of the duality of sign and concept sentences which is the kernel of Wittgenstein's theory of meaning.

There is however a fundamental difference between the construction of arithmetic in a calculus in which the only axioms are function definitions and, say, the postulation of a set of axioms for complex numbers. In the former case the problem of contradiction does not arise acutely because no new definition imposes a restriction on the elements already introduced, but the axiom system for complex numbers deliberately contradicts the familiar properties of real numbers and the question whether the axioms are contradictory is immediately forced upon us. It is not whether complex numbers *exist* which is in question but whether a certain use of signs is consistent or not. It is a confusion of these two situations which on the one hand lead Frege to distrust the formal method and on the other hand gives rise to the misconception that axioms *per se* create concepts. For it is essential to a knowledge of the concept to know its relation to other concepts. Thus the construction of complex algebra as an algebra of ordered pairs is to be preferred to the axiomatic method, not because it is more concrete or 'constructive' but because it reveals the connections (and the differences) between real and complex algebra more clearly.

It has been known for the past 25 years that no formalisation of arithmetic is complete; that there are results which we recognise as correct by stepping out of the formalism which cannot be proved *within* the system. And furthermore that no formalism categorically determines the number concept. This has led, in recent years, to a return to a so called 'realist' view of the nature of mathematics, the view that numbers and their properties belong to the 'real world' and are only inadequately represented by formal systems. Mathematics is once again regarded as a branch of physics studying the world around us, or a branch of psychology studying mental constructs and operations. Important though these discoveries are, they do not constitute a refutation of the view that the concepts of mathematics are created by mathematics. To say that no one formalisation of arithmetic encompasses the whole of arithmetic means only that we recognise arithmetic to be without end, not that the whole of arithmetic already exists in the world around and lies waiting to be discovered. To talk of a 'formalisation of arithmetic' irresistibly tempts us to think that there must be *something* which is being formalised, not just some other 'formalism'. But the fact that no codification is complete or categorical, does not mean that arithmetic is something other than a system of rules. It is often said that Parliament cannot formulate a law which leaves no loophole through which the wily may escape, but this does not lead us to abandon law-making. We continue to make new laws. That no codification is categorical is only another way of saying that no language can be absolutely free from misunderstanding.

REFERENCES

1. Church, Alonzo. 'An unsolvable problem of elementary number theory', *American Journal of Mathematics*, Vol. 58, pp. 345-363.
2. Goodstein, R. L. 'On the restricted ordinal theorem', *Journal of Symbolic Logic*, Vol. 9, No. 2, 1944, pp. 33-41.
3. Hilbert, D., and Bernays, P. *Grundlagen der Mathematik*, Vol. 2. Berlin, 1939.
4. Kleene, S. C. *Introduction to Metamathematics*, Amsterdam, 1952.
5. Markov, A. A. 'On the continuity of constructive functions', *Uspechi Mat. Nauk.*, 1954, pp. 226-230.

VIII

THE AXIOMATIC METHOD

I PROPOSE to talk today about the axiomatic method which is both a method of discovery and itself one of the greatest mathematical discoveries. I shall examine the validity of the method and say something of its fruitfulness and of its limitations in the foundations of mathematics. I shall consider amongst other questions, whether axioms create concepts and how the consistency of axioms is established by the use of models.

I shall start by examining the powerful attack which Frege made on the use of the axiomatic method in the definition of the number concept. The view which Frege so ably sought to overthrow is the view, widely held today, that the problem of the *meaning* of a mathematical concept is totally irrelevant; that it suffices to capture the concept in an axiomatic system. Thus, instead of attempting to discuss the number concept (or any other mathematical concept) we should set up a class of axioms from which all the desired properties of the concept may be derived. This does not, of course, correspond to Euclid's notion of an axiom system, for Euclid considered that the axioms needed to be supplemented by definitions of the concepts. Frege in effect condemned the axiomatic method for separating mathematics from its applications. An axiomatic theory of arithmetic, e.g. a system based on induction and the Peano axioms for the successor function, he pointed out, cannot distinguish between the natural numbers of everyday use and the numbers greater than or equal to ten. And this remains true if we add axioms for addition, though not if we add axioms for multiplication.

It is however quite easy to show that even with multiplication the axioms do not determine the natural numbers uniquely; for instance, the fractions $1/p, p \geqslant 1$, with $1/(p+1)$ as the successor of $1/p$, and addition and multiplication defined by the axioms

$$1/p + 1/q = 1/(p+q)$$
$$(1/p) \cdot (1/q) = 1/pq$$

satisfy induction and the Peano axioms, including recursive axioms for addition and multiplication. In fact Gödel has shown that whatever additional axioms we impose, axioms for subtraction or exponentiation, or what you will, then (provided the axioms are consistent) it is always possible to find a model of the axioms, that is, an interpretation, which is different from the intended interpretation. Frege's standpoint is essentially summed up in his observation that you cannot create the number zero simply by laying down axioms for operations with an oval drawn on paper with a pen. For him, the number zero is a concept, a logical entity, as much a real object as anything apprehended by the senses is real, an object which mathematicians recognise but do not create. Accepting for the moment Frege's criticism of the formal axiomatic method, let us see what he offered in its place. Like the formalists he criticised, Frege himself built up an axiomatic system—one of the most carefully constructed and subtly discriminating systems that had yet been devised, but in Frege's system predicate, not number, is the fundamental concept and number is defined in terms of one-to-one related classes (a class being the extension of a predicate). In this way Frege sought to wed his formal concept of number to the real world, since he could prove for instance that a class like (a, b, c) contains three numbers.

But unless 'class' is taken to have a fixed significance over and above that which the formalism gives to it, Frege's system is as much subject to Gödel's theorem as any other formalism, and Frege's object is not achieved. The *classes* of an axiomatic system are no more real objects, in the ordinary sense of the word, than are the *numbers* of an axiomatic system. Frege's definition of number in terms of classes is a purely technical device, of value and importance in constructing arithmetic in a particular type of axiom system, but no more than that. It is perhaps interesting to observe that the notion of one-to-one correspondence, a notion that has been introduced into mathematics only comparatively recently, is claimed by psychologists to be present in all children before they form the number concept. If we regard counting, as we may, as a process of pattern making (like arranging a group of sticks to make the figure 5), it is as reasonable, from the logical point of view, to establish one-one correspondence between classes by means of counting as it is to define the equality of classes by means of one-one correspondence, and in fact the history of counting, and the strong evidence for the absolute priority of 2-counting, that is counting $1, 2, 2+1, 2+2, 2+2+1, 2+2+2$, and so on, seem to make the view that the concept of a one-one correspondence precedes the number concept, quite untenable. No doubt if Frege had defined correspondence in terms of function, psychologists would by now have discovered the function concept too already present in children. A similar technical device which is currently mistaken for an important fundamental discovery (suitable for teaching to school children) is the reduction of the notion of ordered pair to the class concept by means of the definition of the pair with first element a and second b as the class with elements a and the class (a, b).

Let us return however to Frege's criticism of the axiomatic method. We must concede that Frege was right when he observed that we cannot capture a concept in a system of axioms, for axiom systems are not categorical. Is there then some other way in which mathematical concepts may be defined? The axiomatic method consists in specifying certain properties of the concept from which other properties may be derived; the method is purely formal because no property of the concept which is not specified by the axioms, may be used in the derivation of new properties. When we contrast an axiomatic with an intuitive approach to a problem we are contrasting a derivation from premisses named in advance with a derivation from hidden, i.e. unexpressed premisses. To say that no axiom-system suffices to capture the concept is not to contrast the axiomatic with the intuitive but says, in effect, that the concept *cannot* be made precise. We have the feeling that the concept is somehow in us, only we cannot quite bring it out; but because we are ready to accept, or reject, certain consequences of our assumptions does not mean that we know all possible consequences in advance of drawing them out. To know what one wants is often no more than a readiness to accept or reject what is offered, not to have at hand a model by which the acceptance or rejection is to be made. In the same way if we accept or reject an axiomatisation of a concept because it does, or does not, correspond to our so-called intuitive notion of the concept, we do not necessarily have some other exact notion with which the axiomatisation is to be compared, but perhaps only a disposition to respond to or reject any particular set of axioms. Let me repeat that the fact that axiom systems are not categorical does not, in my view, reveal a deficiency in the axiomatic method, but a want of categoricity in the concept itself.

To put it another way, *nothing can be expressed in a form which is entirely exempt from misunderstanding.*

To gain an insight into the axiomatic method let us consider in detail a particularly simple set of axioms, the axioms for a group. A set G of (at least two) elements is said to form a group with respect to a (not necessarily commutative) operation, symbolised by juxtaposition of elements, if

1. for any elements a, b of G there is an element c such that

$$ab = c;$$

2. for all elements a, b, c of G

$$a(bc) = (ab)c;$$

3. there is a neutral element ϵ in G such that for any a

$$a\epsilon = a; \text{ and}$$

4. to each element a corresponds an inverse element a^* such that

$$aa^* = \epsilon.$$

Do these elements create the concept of a group? Certainly, if we take elementary arithmetic for granted, the axioms have many models, for example, the integers with zero, under addition, zero playing the part of the neutral element, and $-m$ the part of the *inverse*, or just the set of two numbers 0, 1 under addition modulo 2, with 0 again playing the part of ϵ and with $0^* = 0$, $1^* = 1$. By appeal to such models we may satisfy ourselves that the axioms are consistent, or at any rate that they are consistent relative to such a familiar system as the natural numbers, for any contradiction which may be derived from the axioms may also be derived in the model.

Moreover, if we take for granted the concept of a set and the concept of a binary operation on the elements of the set (a mapping of the product set $G \times G$ into G itself), then the axioms do provide us with a criterion for deciding whether or not any particular set and operation form a group, and more cannot be asked of the definition of a concept. In isolating these specific properties of a set in relation to a binary operation, and calling them the group axioms, we see the axiomatic method at its best, and perceive its power to clarify a concept. But we must also recognise its limitations as a method in the foundations of mathematics, for the device we employed to establish the consistency of the axioms is applicable only when the axioms may be confronted by an already established branch of mathematics. As we remarked, we were contented to prove the relative consistency of the axioms and relative consistency presupposes some standard of comparison. Furthermore, the axiomatisation we have been studying (and this is typical of the piecemeal use of the axiomatic method in modern mathematics) is only a partial axiomatisation, since a proper formulation of group theory must rest upon an axiomatic theory of sets, and belief in the consistency of set theory is the triumph of faith over experience.

The third and fourth group axioms have an existential character which is apt to be misleading. We appear to assert the existence of the neutral element and the inverse of an element, as though we could *create* the neutral element of a group by postulating that there shall be one. But in fact an existential postulate is not involved; we are merely describing the characteristics of a group. Thus a set is a group with respect to such-and-such a relation if it has elements with certain properties—the axioms do not create the group concept, but mark out its frontiers.

It is sometimes said that the group axioms create the *pure* concept of group, whereas the groups whose existence ensure the consistency of the axioms are groups contaminated with other properties. Certainly it is true that any structure we care to name, which happens to be a group, has other properties which are not necessarily the essential properties of a group; the elements may be ordered, or cyclic. Two different familiar structures, which are isomorphic as groups (but otherwise retain their identities), are said to be *representations* of the same group, as though there was somewhere the pure group of which the two structures are only representations. As in Plato's theory of ideas the representations appear as mere shadows of the pure group (ignoring the fact that the representations are themselves now only shadows). Even to say, as we have said, that the group axioms delimit the concept of group, may be thought to raise the question of the real existence of a group as opposed to its representations. But these are essentially confusions into which we are led by our forms of expression. If two structures are said to be representations of the same group, when they are group-isomorphic, then to be representations of the same group means just to be group-isomorphic, and the existence problem created by the term 'same group' may be resolved simply by banishing the term; for there is nothing which we can express by means of the language of representations which cannot equally clearly be expressed in terms of group-isomorphism. And the 'pure group' whose existence the axioms may be thought to assert is nothing but the group concept; or if we prefer to avoid the use of concept language, we may say that group is a term which may be eliminated in context by means of the group axioms, that is to say, the expression (S, f) is a group, where S is a set and f is a mapping from $S \times S$ into S, may be replaced by the group axioms enunciated for the particular S and f.

If we seek to apply the axiomatic method in the foundations of mathematics, for instance in the theory of complex numbers, by adjoining to the real number field an element i such that $i^2 + 1 = 0$, we meet the obstacle that the relation $i^2 = -1$ necessarily denies the fundamental property of real numbers that every square is non-negative. We are not now listing common properties of an already familiar structure but instead we are seeking to postulate the existence of a structure with properties which conflict with the properties of all structures known hitherto. It is this acute need to prove that the adjunction of the new element i does not introduce a contradiction into the field axioms (as in fact it does contradict the order axioms) which gives *constructive* methods their great importance in foundation studies. As is well known, several constructive courses are open to us; we may set up the complex numbers as an arithmetic of ordered pairs of real numbers, or following Cauchy, as the field

I

of remainders in the ring of polynomials over the ordered real field, *modulo* $x^2 + 1$. The second method has the advantage of generating the complex field by a general process for generating fields; the method of ordered pairs is conceptually simpler but may seem rather artificial unless it is seen as one step in a chain of constructions with ordered pairs. The interesting question however is what *is* a complex number? Is the ordered pair just one representation of the complex number, the complex number itself being created by the act of adjoining an element i, with $i^2 + 1 = 0$, to the real field? But if the act of adjoining i to the real field were genuinely a creative act, then no question of consistency would arise. I think we must either reject the process of adjoining an element to a field as nonsensical, or regard it merely as an abbreviated version of Cauchy's constructive generation of a field by taking remainders in a polynomial ring.

Cauchy's method, moreover, constructs complex numbers as polynomials of the first degree, and a polynomial of the first degree is none other than an ordered pair; as for polynomials, a polynomial is either an ordered set of elements of a given field F, or is an element $a_n x^n + a_{n-1} x^{n-1} + \ldots + a_1 x + a_0$ of a super-field F^+ which contains x, and the field F of the coefficients as a sub-field, and if a super-field is at our disposal, what point is there in seeking to create it by adjoining the element x to the field F? Thus it transpires that whatever course we adopt, we construct the complex number as an ordered pair of real numbers, and the choice of the route by which this construction is made is irrelevant.

Of course the definition of a complex number as a pair of real numbers presupposes the existence of real numbers, and the problem of defining real numbers is much more difficult. There is no difficulty in presenting a set of axioms for the real numbers; we have only to add suitable axioms of order and completeness to the field axioms. The whole difficulty lies in proving the consistency of the axioms. None of the familiar elementary systems, natural numbers, integers, rationals or algebraic numbers, satisfies the completeness axiom which requires the existence of a limit for every bounded monotonic sequence. Several so-called constructive theories of real numbers are known; in Cantor's theory real numbers are defined as convergent sequences of rationals, and in Dedekind's theory as certain classes of rationals. Both these theories require set theory for their complete formalisation and as we have already had occasion to observe, the consistency of set theory rests on faith, not reason. In recent years attempts have been made to construct fragments of real number theory without general set theory, by considering for instance recursive sequences of rationals, or recursive sets of rationals; if we confine ourselves to primitive recursive sequences and sets the Cantor and Dedekind theories cease to be equivalent, and even if we accept general recursive sequences and sets the equivalence of the two theories can only be proved by nonconstructive methods.

If we continue to retrace our steps from real numbers to rationals, integers and finally to natural numbers, there is no serious existence problem until the natural numbers are reached, for integers and rationals may readily be constructed from finite ordered sets of natural numbers. Natural numbers may be defined within set theory, in various ways, as Frege and Russell showed, but if we seek to place number theory on a surer foundation than the theory of sets, we meet the existence problem in its most acute form.

I want to start considering this problem by contrasting two axiomatisations of the natural numbers. We may characterise the system N of natural numbers (without zero) by the following five axioms:

1. There are two mappings of pairs of elements of N into N, denoted by $m+n$ and $m \cdot n$ respectively which are associative, commutative and distributive (the second over the first);
2. N has a member 1 such that $m \cdot 1 = m$ for all m of N;
3. if either $m+x=n+x$ or $m \cdot x=n \cdot x$ then $m=n$;
4. for any two different members m, n of N there is an x in N such that either $m+x=n$ or $m=n+x$;
5. any property of the number 1 which is a property of $n+1$ when it is a property of n, is a property of all members of N.

The first criticism which must be made of these axioms is of their sheer existential character; we cannot now turn the axioms round and say that any system which has the properties 1-5 is a system of natural numbers; since we are by hypothesis taking the first step in system construction, we have no model to which appeal may be made to secure consistency. What sense then shall we attach to the assertion of the existence of an element 1 such that $m . 1 = m$? Does the mere assertion of existence create the number? And what of the mappings whose existence are affirmed in the first axiom? Is mapping here a mathematical concept, and if so how is it defined? The axioms do not help us. Perhaps the axioms are not really about mappings but about the use of the signs '+' and '.', and if this is so then axioms 1 and 2 do not assert existence, but tell us that if 'm', 'n' are numerals then '$m + n$' and '$m . n$' are numerals, and that '1' is a numeral, from which it follows that, for instance, '$1 + 1$', '$1 + 1 + 1$', . ., '$1 . 1$', '$(1 + 1) . (1 + 1 + 1)$', . . are numerals. Assuming that the relation of equality is to be taken as the relation of identity between numerals, and that the axioms are consistent unless they entail $1 + 1 = 1$, the relations asserted in axiom 1, that '$m + n = n + m$', '$m . n = n . m$' and '$l . (m + n) = l . m + l . n$' for any numerals '$l$', '$m$', '$n$' once again raise the problem of consistency. It is the function of these relations to transform such a numeral as '$(1 + 1) . (1 + 1 + 1)$' to the standard form '$1 + 1 + 1 + 1 + 1 + 1$', but how can we tell that using the relations in different orders will not affect the reduction to standard form, and so by axiom 4 lead to an equation like $1 = 1 + 1$? What is at issue is the uniqueness of the standard form; the axioms fail to make uniqueness evident and (lacking arithmetic) we have no means at our disposal for proving it.

By talking of numerals instead of numbers we free the axioms from their existential guise, but are we not tacitly affirming the existence of entities (the numbers themselves) of which numerals are the signs? Surely, we may say with Frege, the subject of arithmetic is number not numeral.

But is this not just a question of language, whether we should use object or concept language? Certainly, when we say that the number 2 is smaller than the number 5 we are not saying something about the sizes of the numerals '2' and '5', but nevertheless we can express the order relationship of the numbers 2 and 5 without making any reference to numbers, for instance, by saying that the numeral '5' is a transform of the numeral '$2 + 3$'.

Frege sought to explain the equation

$$5 + 2 = 3 + 4$$

by saying that the numerals '$5 + 2$' and '$3 + 4$' have different senses but the same denotation, and was then obliged to find something (namely, the number 7) which the signs denoted. Occam's razor encourages us to find an explanation of the equality which does not introduce additional concepts, and such an explanation is to hand in the theory of sign transformations.

A very different set of axioms for the natural numbers were given by Peano; these consist of the axioms

$$0 \neq Sx,$$
$$(x = y) \longleftrightarrow (Sx = Sy),$$
$$x + 0 = x,$$
$$x + Sy = S(x + y),$$
$$x \cdot 0 = x,$$
$$x \cdot Sy = x \cdot y + x$$

together with induction. In the Peano axioms the risk of contradiction is lessened, since addition and multiplication are now defined constructively and do not impose additional connections in an established structure, but the problem of consistency remains. We can however go much further than Peano, and construct arithmetic in a calculus in which *all* axioms are replaced by function definitions, as in my equation calculus, and the risk of contradiction is reduced to a minimum.

Few discoveries in mathematics are comparable in interest and importance with the discovery of a fruitful set of axioms, and if the axioms are not only fruitful, but simple and few in number, and these few demonstrably independent, then the aesthetic appeal of the discovery is enormously enhanced. But the usefulness of the axiomatic method is limited by the need to prove consistency; as Gödel's second incompleteness theorem shows, no axiom system can be proved consistent by methods formalisable within the system itself, and so the nearer one approaches to the *foundations* of mathematics (be it arithmetic or the theory of sets) the less reliance can be placed upon the axiomatic method. At the very threshold of mathematics the usefulness of the method vanishes entirely and the discovery of new constructive methods becomes of paramount importance.

IX

PURE AND APPLIED MATHEMATICS

THE subject of this paper is not the differences in content between works of pure and applied mathematicians, not the differences in their respective attitudes to their subject, however important these may be, but the problem of the very possibility of applying mathematics to the real world, and of the means by which this application is made.

On one view of the foundations of mathematics pure mathematics is just a game like chess, an immense and miraculously wonderful game, but just a game for all that. Certainly chess has much in common with mathematics; chess is no mere repetition of familiar moves, like noughts and crosses, no mere mechanical traverse of a well-worn path, but a field for the exercise of imagination and inventiveness of a very high order. But mathematics is more than *playing* a game, even a game of such fertility as chess; it is the continual creation of new games and the comparative anatomy of games. Wittgenstein looked to the application of mathematics to distinguish it from a mere game, but mathematics does not need applications outside itself to mark this distinction, except in the sense of game in which every human activity like painting, creating music or writing books is a game. In this sense physics and chemistry are also games when they are not technologies.

At the other extremes we meet the view, a view which still has its defenders, that even the laws of logic are laws of nature, namely the laws of mental behaviour. On this view we accept the *tertium non datur* or any other law of logic, because we are forced to accept it by our very natures, and had our brains been differently constituted then the laws of logic would also have been different. To the extent that there are racial characteristics in the forms of discussion and debate this view correctly represents the origins of logic, but it takes no account of the evolution of language. This evolution is clearly marked in the development in mathematics of the concept of a formal system. I shall examine this development in some detail because it throws light not only on the nature of pure mathematics but also on the question with which I am chiefly concerned, the question of the possibility of applying mathematics and of the means of application.

Mathematics started (or had one of its origins) we may suppose in the practical geometry of the Egyptian builders and surveyors. The concept of equal inclinations and right-angle grew from observation of a plumb line, and the need for constructing a right angle in a horizontal plane lead to the discovery of the 3, 4, 5 triangle and the angle in a semi-circle. A people with the insight to make such discoveries certainly has some form of mathematics even if it leaves no systematic record. The Greeks took over this drawing-board geometry, rapidly increased the reservoir of known constructions and then made a decisive step forward in the evolution of mathematics: *they began to relate constructions one to another.* The construction for drawing an equal sided triangle was related to the construction for dividing a line into two equal parts, and this in turn was related to the problem of finding the centre of a circle. The amount of information you need to have about a triangle before it can be constructed must early have become apparent and we can imagine the interest which the diversity of this necessary information aroused. Once it was known that a triangle can be constructed when we are given three sides, or two sides and an included angle, or two angles and a side, the natural development was to recognize that *triangles which share these minimum construction properties are equal in all respects,* and this marks the

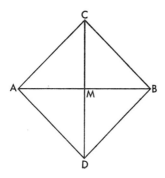

beginning of mathematical proof. For example the *construction* for bisecting a line AB by drawing two equilateral triangles ABC, ABD on AB and then joining C to D to cross AB at its mid-point M, may be seen, in turn, as a *construction* for two triangles ACD, BCD which have all their corresponding sides equal, and therefore also their angles equal, and a *construction* for two triangles ACM, BCM which have two corresponding sides and their included angles equal, making AM and BM equal. At this stage in the evolution of the concept, proof consists in breaking down constructions into constituent constructions; formal logic has no part to play. Even Pythagoras's Theorem might have been proved by cutting up and rearranging a figure (although there is no historical evidence for supposing this proof is much earlier than A.D. 900). It is important to notice that there are two ways in which a proof by cutting and rearranging may be conceived; we may rely on the *accuracy* of the drawing and cutting to show that certain parts of a figure are equal or we may use cutting and rearranging to *draw attention* to certain equalities which *result* from the construction. It is only this latter use of cutting up a figure which is a stage in the evolution of the concept of proof—the former use suggests connections to look for, but otherwise does not belong to pure mathematics.

The next stage in development is a concise enumeration of the basic constructions to which *all* constructions may be broken down. The choice of basic constructions is to a certain extent arbitrary, and becomes important only when a principle of economy begins to operate, and we seek for the shortest list of basic constructions. At this stage the notion of an axiom in the modern sense begins to emerge; attention is focused not on the initial constructions as constructions, but on the relation of the initial to the subsequent constructions. Before this stage is reached it makes no sense to talk of dispensing with the *drawing instruments*, for the whole of geometry is still drawing, but once we have isolated the initial constructions, geometry becomes the study of the relationships of these constructions to other constructions and the total loss or destruction of the instruments leaves these relationships untouched. We may forget the very meanings of the initial constructions, or find other meanings; the initial constructions have become, not constructions, but substitutes for constructions. In the study of these relationships formal logic plays little or no part; perhaps one or two *common notions*, like the transitivity of equality, are made explicit, but the underlying logic remains intuitive, that is to say it can be found only by studying proofs as natural phenomena. This stage of mathematical development lasted until about 1870 and recurred in the second and third decades of the present century when the intuitionists were unable or unwilling to codify their logic and those outside the intuitionist circle could determine what constituted a valid proof in intuitionist mathematics only by analysing acceptable proofs—intuitionistic logic was to be known only by its fruits.

The end of this stage of mathematics began with Frege's great work. Even if Frege was largely mistaken in his views of the nature of mathematics, by his failure to recognise the relativity of mathematical concepts in his search for an absolute logic of mathematics, nevertheless his programme to make explicit the rules of proof marks the opening of the modern era. In the penultimate stage of this evolution mathematics and logic are entirely divorced from physics and psychology. Mathematics is conceived of as a purely formal calculus, a game with initial positions and rules for moving the pieces, an uninterpreted calculus; and the centre of interest passes from mathematics to *metamathematics* conceived of as the comparative study of formal systems by means once again of an intuitive logic. Like a snake swallowing its tail, intuitive logic operates on its own codification. But metamathematics itself of course is formalisable and the final stage is perhaps an infinite hierarchy of formal systems.

Two discoveries were critical in the evolution of the formal system. The first of these, as is well known, was the construction of models of non-Euclidean geometry, within Euclidean geometry itself. The consistency of Euclidean geometry was thought to be guaranteed by the physical model of drawing-board geometry out of which Euclidean geometry grew; but this drawing-board geometry conflicted with such properties of non-Euclidean geometry as the angle sum of a triangle, and so the validity of the drawing-board model as a test of consistency became suspect, and geometry was set free from its earthy roots. Whether the geometrical objects are 'things' which satisfy the axioms, or whether the axioms create the geometrical objects now becomes a subject of dispute. Some maintained, and this view is still widely held by mathematicians, that the axioms are only the mould from which the geometrical objects may be made; that for instance the Euclidean triangle is any mathematical object, such as a set of numbers, which satisfies the axioms. The difficulty in accepting this position is that it leaves open the question how these objects are themselves defined; if they too are defined by axioms, then we enter a vicious circle, and if not, then their nature remains to be determined. There are several roads from this point. It could be argued that mathematical objects are objects of the real world, like tables and chairs, but this appears to confuse the triangle of geometry with its ancestor the triangle of the drawing board. And if we say that the triangle of geometry is not exactly this, but an idealisation of it, this is just to say that the triangle of geometry is *not* a real object. And if we say that it is a mental structure, then we are again making a confusion, this time between the triangle of geometry and a mental image of this triangle. Frege, rightly, called the triangle of geometry a *logical* concept, but he might as well have said a *mathematical* concept for all that it helps us. The opposite view, that the axioms create the concept, is also open to strong objections. In the first place axiom systems are neither unique nor categorical. There are, for instance, various axiomatisations of Boolean algebra, and so far from characterising one clear concept these axioms admit as models such diverse objects as the subclasses of a given class or the divisors of the number 30. The very expression 'axiomatisations of X' leads us to look for X outside axiom systems. We hold exhibitions of pictures, but pictures are painted outside exhibitions. We formalise mathematics in axiom systems but mathematics is created outside these systems. What then are mathematical objects? The importance of looking at axiom systems in seeking an answer to this question is shown by the fact that there is not one unique triangle concept in mathematics. The Euclidean triangle is different from the Lobachevsky triangle, and each of these is different from the Riemann triangle. To detect these differences we must look at their respective properties, and these properties are contained in the theorems of the different axiomatic systems associated with the names of Euclid, Lobachevsky and Riemann. It may well be true that an axiom system fails to do what it is set up to do, that is to say, fails to circumscribe the concepts it is intended to delimit, and in fact as we know from Skolem's work, every axiom system necessarily admits of a second, unintended, interpretation, but this does not mean that we must look for the concept *outside* mathematics, only that no single formalisation encompasses the whole of mathematics. This is hardly surprising when we reflect that no single formal system can define all real numbers, that in a certain technical sense there must be more real numbers undefinable than definable in any system. Of course these limitations relate to certain current types of formalisation and are therefore relative and not absolute, but we know of no formal structure to which they do not apply.

If axiomatic systems are not categorical and if we admit no other source of the concept than the theorems of the systems then we are forced to admit that the concept itself is not sharply delimited. This is a point to which I shall return later, but I want first to discuss in more detail the view that we have an intuitive knowledge of mathematics which only finds partial expression in a formal system. This is akin to saying that we often have ideas which we cannot express in words, meaning that we have certain mental pictures, or tendencies or hopes which are not satisfied by any sentence we formulate. But such shadowy musings surely cannot provide a foundation for mathematics. And even the conviction that mathematicians so often have that they are discovering an order or pattern that is already present in the world, and not inventing it, is more related to the psychological accompaniments of research than to its results; that is to say this conviction arises from the way mathematical discovery is made, the strange jumps in the dark which take place, the feeling of certainty that so often anticipates proof (but is perhaps as often a prelude to disappointment and failure). All these are elements in the psychology of reasoning and are not to be confused with mathematics itself.

If instead of regarding a formal system as an imperfect expression of intuitive mathematics, we regard formal systems as quasi-physical objects of study by intuitive mathematics (now called metamathematics), then (from the logical point of view) we are back again to one of the early stages in the evolution of mathematics, with formal systems and their syntax taking the place of geometrical constructions. We have in fact corroborative evidence that mathematics repeats stages in its development in the parallels of the nineteenth-century theory of real numbers to the Eudoxus theory of proportion, and the so-called paradoxes of Zeno to the contemporary rejection of intuitionism. The essential point I want to make is that *whether proof procedures are made explicit and rigid in a formal system, or left fluid and implicit in a so-called (and ill-called) intuitive presentation the nature of mathematical objects is the same.* A mathematical object is defined by, and only by, the theorems which are proved about it, whether the body of theorems constitutes the deduction of a single axiom system, or an uncodified family of relationships.

We are now in a position to answer the question of the possibility of applying mathematics so far as it relates to the applications of geometry in Engineering and Architecture or terrestrial physics. *In its application geometry retraces the course of its evolution from the drawing board to the formal system.* Geometrical drawing reverts to its dual role of construction in physical space and model of logical relationships. The drawn circle and the drawn line meet in two points, the drawn angle in a semicircle *is* a right angle in physical space (so that two such right angles differ imperceptibly from a straight line) and so on. If physical space were to change so that these coincidences no longer occurred, Euclidean geometry would lose its application, but not its logical validity. To sum up, Euclidean geometry is applicable because it grew out of observation and experiment, the expressions of observed regularities (two points lie on only one stretched thread, etc.) passing in the course of time from statements of fact into conventions, and so into axioms in a formal system; geometry is applicable because the world has not changed since geometry evolved from the drawing board.

This account of the applicability of geometry may seem at first sight to conflict with the views of modern physics that Euclidean geometry is not the geometry of physical space. If the application of geometry is possible only because geometry grew out of geometrical drawing how can a non-Euclidean geometry, which conflicts with the geometry of the drawing board, find an application? Of course the facts of the drawing board—the observable facts—tell us nothing of the physical properties of very large triangles (whose sides are light paths between planets). The physical geometry of very large triangles is virtually unknown to us. Now it may transpire—and this would be no more than a remarkable coincidence—that the physical geometry of very large triangles has the same relationship to a non-Euclidean geometry that drawing-board geometry has to Euclidean geometry. That is to say, the physical geometry of large triangles may show certain regularities (e.g. in the manner of variation of the angle sum for triangles) and these regularities may correspond to the axioms of a certain geometry, so that this geometry would then admit an application to large triangles in space.

It must be remarked, too, that the observed regularities of drawing-board geometry, and therefore the evolution of Euclidean geometry is in a sense the outcome of the *imperfections* of our vision and of our drawing equipment. For had we seen the crossing of two lines, not as a point but as a family of points, and had we recorded angle sums as families of numbers, Euclidean geometry might never have been conceived. It is a familiar experience that the research worker can see too much as well as too little.

If we turn from geometry to Newtonian mechanics we meet a branch of mathematics in a different state of development. Although particle mechanics has long been formulated more or less satisfactorily as an axiom system, the axiomatisation of the mechanics of continuous media has only just begun. The nineteenth-century illusion that the mechanics of continuous media is a by-product of the mechanics of particles died hard, but recent work on elastic and plastic materials has finally dispelled it. The application of mechanics presents no new problem; the interest of mechanics from the viewpoint of the study of the nature of mathematics lies in the concepts of mechanics, force, mass, velocity, acceleration and energy. Not all of these are quantities which are directly measurable—for instance mass is measured indirectly in terms of weight—and we see that a mathematical system may be applicable so to speak only at certain points. We are inclined to suppose that mechanics and geometry are fundamentally different because mechanics can make predictions but geometry cannot. Mechanics predicts an eclipse or the existence of an undetected planet, but what does geometry predict? Well, that if we perform such and such a construction then a certain three lines will meet at a point; such predictions may not be very exciting but, and this is the important point, looking at geometry in this way shows us what the nature of prediction in mechanics, or for that matter in physics, really is.

I turn now to arithmetic, the oldest branch of mathematics, and the one which reveals the problem of the application of mathematics most clearly. Unlike the case of geometry we know nothing of the origins of arithmetic, but it is reasonable to suppose that these two branches of mathematics had similar evolutions. The arithmetical *theorem* that $2+3=5$ grew out of repeated experiments in counting, counting stones, or apples or sunsets. The first stage in the evolution of the number concept must have been naming collections, two, three, four, and perhaps a few more names. Doubtless this first stage was preceded by a time in which small collections of particular objects had names, like a brace of dogs, but it is the naming of collections as collections which marks the birth of the number concept. At this stage numbers are not related to one another, but are names given to objects of immediate perception like

.   .   .   .   .

Experiment now shows that, for instance, two and three make five, and in this way various relationships between numbers are discovered, like two and one make three, three and one make four, four and one make five, and when these facts have been assimilated we are ready for the decisive step, the discovery of counting. Counting is a process of calculation, which replaces the immediate correlation of a name with a collection by a step-by-step determination of the name, using the relationships 'one and one is two', 'two and one is three', 'three and one is four', 'four and one is five' and so on. It is commonly supposed that counting establishes a one-to-one correspondence between the numbers and the elements of the collection counted, but it

seems to me that what counting does is to name in turn subclasses of the collection to be counted, passing from subclass to
subclass by the addition of a single element at a time. I can
support this view of the nature of counting on two grounds. On
the one-to-one relation theory of counting, the number of a
collection is the last number named; thus when counting five
dots

$$\cdot \quad \cdot \quad \cdot \quad \cdot \quad \cdot$$
$$\text{I} \quad 2 \quad 3 \quad 4 \quad 5$$

we write in turn the numbers 1, 2, 3, 4, 5 beneath the dots, the
last figure 5 naming the whole collection; but if this is so then
necessarily the figures 2, 3 and 4 name the subcollections they
terminate, and so even on the one-to-one correspondence
theory, we are in fact counting subsets. Since this is so, and since
counting the subsets explains counting without recourse to the
additional concept of the one-to-one correspondence, it follows
that this additional concept is dispensable and superfluous for
understanding the counting process. Another reason I have for
this view is that some number signs appear to be patterns
formed of the elements of the collection they number (or
degenerate forms of these patterns) like the signs

$$\text{\Large 5} \quad \text{and} \quad \text{\Large 6}$$

which are made of five and six sticks respectively; if we invent
similar signs for the numbers two, three and four, for instance

$$\text{\Large ⅃} \quad \text{\Large ⊐} \quad \text{\Large ⅃}$$

we can describe counting sticks as a process of arranging them
into patterns. The patterns I have chosen are such that faced
with a pile of sticks, we form in turn the patterns

$$| \quad \text{⅃} \quad \text{⊐} \quad \text{ᄂ} \quad \text{⌐}$$

by adding a stick at a time, and the pattern formed when all the sticks have been used up is the number of the collection. It will be observed that what we are doing is to number the sub-collections in turn. On this theory of number signs we certainly do not associate a number with each element of the collection but only with the successive subsets. What I think has misled mathematicians into thinking that counting is a one-to-one corre-lation of numbers and objects is the familiar recital of the words *one, two, three* and so on, when we count. But it seems to me to be more likely that we are reciting not the numbers, but the number relations 'one and one is two', 'two and one is three', 'three and one is four', and so on, where each 'and one' is not said in words but by pointing to the object counted. Another way of describing this explanation of counting is that in count-ing we first *regard the collection as a number sign,* and then trans-form this sign into a standard form, taking the familiar names two, three, four and so on, as standard forms, or standard names of the numbers.

At the next stage in the evolution of arithmetic the observed relationships, like $2 + 3 = 5$, become accepted conventions, and the scene is set for an axiomatic treatment, which deduces these relationships as theorems from a certain set of postulated rela-tions. The axiomatisation of arithmetic came remarkably late in the history of the subject, Dedekind's (1870) recursive definitions of sum and product being probably the earliest attempt at formalisation, and Frege's the next.

The dispute whether number is a class of classes or a primitive concept, is not as was once thought a dispute whether numbers are 'real' or 'formal' objects. The classes of an axiomatic theory of classes are no more the collections of everyday experience than the numbers of formalised arithmetic are physical objects. Whether or not numbers are classes of classes depends upon the logic we are using; in a logic which admits quantification over classes we may use classes of classes as numbers, but in a lower order logic, numbers are necessarily primitive elements. Which-ever definition we adopt, no formalisation of arithmetic can exclude a second unintended interpretation of the number variables, but as we remarked earlier, this does not make number a non-formal concept. We may recognise that a certain system does not fulfil our intentions without there being some object in the world or in our thoughts with which we compare the formal system to discover its deficiencies.

In the application of arithmetic, to commerce or to physics, we retrace the evolutionary development, replacing theorems by statements of observation, and so long as these observed regularities persist arithmetic will be capable of application. In the sense in which we spoke of geometrical predictions we can make arithmetical predictions of the kind that when the contents of two bags of 20 apples are counted we shall find 40 apples. In the application of arithmetic some arithmetical relations play a double role; if we buy three boxes of chocolates at 4s. each box and receive 8s. change from a pound note, the calculation $20 - 3 \times 4 = 8$ which we perform belongs to formal arithmetic, but counting the boxes of chocolates and counting the change are the means by which we apply arithmetic to the real world, the statement $20 - 3 \times 4 = 8$ playing a dual role as a theorem of arithmetic and as prediction about the world.

Of course we cannot be certain that a prediction will come true. If the structure of the world changed arithmetic would lose its application, for the certainty of mathematics arises from its divorce from reality. And even if we seek to replace a positive prediction by a probability assertion this too must jump the gulf between convention and reality. The theory of probability like arithmetic or geometry is born of our preception of regularity, then schematised and axiomatised, and finds its application in retracing its evolution. To say of some event that its chance of occurrence is $3/4$ is to say that the future will be like the past just as dogmatically as to say that the sun will rise tomorrow.

I have sought to distinguish mathematics from the real world by contrasting convention with experiment, but the difference is not as precise as this may sound. There are concepts outside mathematics as formal as mathematics itself as we see in certain codes of behaviour and in some branches of jurisprudence, like the law of corporations, where by *formal* I mean not subject to doubt. And we could make any concept formal in this sense by arbitrarily shapening its definition, but this would not extend the range of mathematics. On the other hand mathematics itself may admit non-formal notions; the concept of constructive function is such a notion and mathematicians are by no means all

K

agreed as to what functions are constructive and what are not. And class is another example. But though the boundary is not a sharp one the distinction between mathematics and the world remains and the application of mathematics is to be understood it seems to me only in terms of the dual role of mathematical propositions which I have described.

X

THE DECISION PROBLEM

THERE are many questions in mathematics to which an answer can be found by a purely mechanical procedure. For instance we can determine the valid statements of the class of the statements

n is prime

by the sieve of Eratosthenes; the numbers 2, 3, 4 and so on up to any assigned n are written in a row, and we begin by striking out every second number, starting with 2. Of those which are left, the first is 3 and 3 is prime. Starting with 3, we then strike out every third number, and the first of those which remain is 5, which is prime. Starting with 5 we strike out every fifth number, determining the prime 7, and so on.

Such a mechanical procedure for deciding the validity of a class of sentences is called a *decision procedure* for the class. The *decision problem* for a class of sentences is the problem of finding whether or not a decision procedure for the class exists, and exhibiting the procedure when there is one. If a decision procedure for a class is known, the class is said to be *decidable*; if on the other hand it can be shown that no decision procedure is possible, then the class is said to be *undecidable*.

Of course the decision problem is of interest only if the class of sentences concerned is sufficiently rich; for instance when the class comprises all the statements of some branch of mathematics, or ideally the whole of mathematics.

The branches of mathematics for which the decision problem has been solved are characterised, not only by their mathematical but also by their logical content. An account of the decision problem must therefore be prefaced by some remarks, necessarily brief, about the concepts and notation of symbolic logic.

The basis of symbolic logic is sentence logic which studies the *combinations* of sentences which are valid independently of the validity of the constituent sentences. The elements of sentence logic are sentence variables p, q, r, . . .; negation '$\rightharpoondown$', and the logical connectives 'and', 'or' and 'implies' denoted by '&', ' v ' and ' $\rightarrow$ ' respectively. A sentence is either a sentence variable or has one of the forms $\rightharpoondown S$, S & T, S v T, S $\rightarrow$ T where S, T are sentences. Sentence logic may be formulated either as an axiomatic theory or by means of truth tables.

If we take '$p \rightarrow q$' as an abbreviation for '$\rightharpoondown p$ v q' and 'p & q' as an abbreviation for $\rightharpoondown(\rightharpoondown p$ v $\rightharpoondown q)$, sentence logic may be based on the four axioms

$$(p \vee p) \rightarrow p, \quad p \rightarrow (p \vee q), \quad (p \vee q) \rightarrow (q \vee p), \quad (p \rightarrow q) \rightarrow (r \vee p \rightarrow r \vee q)$$

and the inference rule

$$\frac{\begin{array}{c} S \\ S \rightarrow T \end{array}}{T}$$

permitting the derivation of T from proved sentences S and S $\rightarrow$ T.

A proof in sentence logic is a finite sequence of sentences each of which is either an axiom, or is derived from an axiom by substituting sentences for sentence variables, or is derived from previous sentences of the proof by the rule of inference.

Calculation with & and v may readily be shown to follow the associative, commutative and distributive laws (like + and . in arithmetic). Writing p for $\rightharpoondown \rightharpoondown p$, $\rightharpoondown p$ v $\rightharpoondown q$ for $\rightharpoondown(p$ & $q)$, $\rightharpoondown p$ & $\rightharpoondown q$ for $\rightharpoondown(p$ v $q)$ and $\rightharpoondown p$ v q for $p \rightarrow q$ we may transform any sentence into an equivalent sentence in conjunctive normal form

$$\phi_1 \text{ \& } \phi_2 \text{ \& } \ldots \text{ \& } \phi_n$$

where each ϕ_i is a disjunction of variables or negated variables. By transforming $\rightharpoondown S$ into conjunctive normal form S may be brought into the disjunctive normal form

$$\psi_1 \vee \psi_2 \vee \ldots \vee \psi_n$$

where each ψ_i is a conjunction of variables or negated variables.

Instead of postulating axioms we may define the logical operations by so-called truth tables.

The truth tables for negation and the connectives are as follows, T and F standing for true and false respectively.

→	p	&	v	→	q
F	T	T	T	T	T
F	T	F	T	F	F
T	F	F	T	T	T
T	F	F	F	T	F

To find for example the value of p & q for assigned values of p and q we look in the column headed & and in the row in which the assigned values of p and q occur. In the third row for instance we see that p & q has the value false if p is false and q is true.

It may be shown that all provable sentences are universally valid, i.e. take the value T for all values of their variables, and conversely that every universally valid sentence is provable, so that the axiom system is *complete* with respect to the truth tables.

Sentence logic is itself an excellent example of a decidable system, the truth tables providing a decision procedure. For instance to test whether the sentence

$$\neg(p \to q) \to (p \,\&\, \neg q)$$

is universally valid or not we calculate the value T or F of the sentence for each assignment of values T, F for the variables p, q, the calculation being set out as follows:

→	(p	→	q)	→	(p	&	—	q)
F	T	T	T	T	T	F	F	T
T	T	F	F	T	T	T	T	F
F	F	T	T	T	F	F	F	T
F	F	T	F	T	F	F	T	F

The value T is obtained (in column five) for all values of p and q so that the sentence is universally valid.

It is quite easy to give an even simpler decision procedure for sentence logic. If we take the values of the sentence variables to be o, i (the former standing for truth and the latter falsity) and if we write $1-p$, $p+q$, $p \cdot q$, $(1-p)q$ for $\neg p$, p & q, $p \lor q$ and $p \rightarrow q$ respectively, then to test the validity of a sentence we have only to work out the value of its representation to find if it is zero or not. The sentence considered above is represented by the formula

$$[1 - \{1 - (1-p)q\}]\{p + (1-q)\}.$$

If this takes the value unity, each factor must be a unit and so

$$p + (1-q) = 1$$
$$(1-p)q = 1;$$

from the second we see that $q = 1$, $p = 0$ and this contradicts the first, showing again that the sentence is universally valid.

The conjunctive normal form provides yet another decision procedure. The given sentence

$$\neg(p \rightarrow q) \rightarrow (p \text{ & } \neg q)$$

may be transformed into

$$\neg\neg(\neg p \lor q) \lor (p \text{ & } \neg q)$$

and thence into $(\neg p \lor q) \lor (p \text{ & } \neg q)$

and finally into $(\neg p \lor q \lor p) \text{ & } (\neg p \lor q \lor \neg q)$,

and since each bracket contains a variable and its negation the sentence is recognised as universally valid.

In sentence logic we operate upon sentences as unanalysed wholes. In the next stage in the development of logic we proceed to analyse sentences into subject and predicate. In predicate logic, as this second stage is called, we distinguish three classes of variables

(i) Sentence variables p, q, r, . . .
(ii) Individual variables x, y, z, . . .
and   (iii) Predicate variables $P(x)$, $Q(x,y)$, $R(x,y,z)$, . . .

Predicates may be valid for all values of their arguments or for some, or for none. If a predicate $P(x)$ is valid for all values of its argument we write

$$AxP(x)$$

(read 'for all x, $P(x)$'), and if a predicate $P(x)$ is valid for some values of its argument we write

$$E x P(x)$$

(read 'there is an x such that $P(x)$').

The operators A, E are called quantifiers; A is the universal and E the particular or existential quantifier. A sentence of predicate logic is either a sentence variable, or a predicate variable, or a combination of sentences $\rightarrow$S, S & T, S v T, S $\rightarrow$T or takes one of the forms $A x P(x)$, $E x Q(x)$.

As axioms for predicate logic we may take

$$A x P(x) \rightarrow P(y), \quad P(y) \rightarrow E x P(x)$$

and for drawing inferences the rules

$$\frac{S \rightarrow P(x)}{S \rightarrow A x P(x)}, \quad \frac{P(x) \rightarrow S}{E x P(x) \rightarrow S}$$

(where x is not a free variable in S) reaffirming the rule

$$\frac{S}{\quad S \rightarrow T \quad}{T},$$

the sentence below the line being inferred from that (those) above by the rule in question.

Taking for granted some obvious rules about changing variables we may say that a predicate sentence is provable if it is an axiom, or a valid sentence of sentence logic, or is derived from a provable sentence by substituting sentences for sentence variables or is derived from provable sentences by the rules of inference. An important property of both predicate logic and sentence logic is the *deduction theorem* which says that if a sentence T is derivable from a hypothesis S (no substitution in the variables in S occurring in the derivation) then S $\rightarrow$T is a provable sentence.

As an example of the use of the deduction theorem we prove the sentence

$$(A x)\{P(x) \rightarrow Q(x)\} \rightarrow \{(E x)P(x) \rightarrow (E x)Q(x)\}.$$

From the hypothesis $(A x)\{P(x) \rightarrow Q(x)\}$

and the axiom $(A x)\{P(x) \rightarrow Q(x)\} \rightarrow \{P(y) \rightarrow Q(y)\}$

we derive $P(y) \rightarrow Q(y)$

and hence, by the axiom $Q(y) \rightarrow (Ex)Q(x)$

we derive $P(y) \rightarrow (Ex)Q(x)$

from which we infer $(Ex)P(x) \rightarrow (Ex)Q(x)$

and by the deduction theorem conclude

$$(Ax)\{P(x) \rightarrow Q(x)\} \rightarrow \{(Ex)P(x) \rightarrow (Ex)Q(x)\},$$

as desired.

To provide a content for the sentences of predicate logic we suppose given some domain of individuals to which the individual variables and the quantifiers A and E refer. A value of a predicate variable is simply an assignment of a truth value for each assignment of individuals from the domain for the individual variables in the predicate. For instance in a domain of two individuals 0, 1 there are just 4 predicates with one argument, 16 with two arguments, 256 with three arguments; each predicate with two arguments is given by a table like

	0	1
0	T	F
1	F	T

The predicate given by this table, $P(x, y)$, say, is such that $P(0, 0)$ is true, $P(0, 1)$ is false, etc. The sixteen ways of filling in the T's and F's in the table give the sixteen possible predicates with two arguments.

A sentence in predicate logic is said to be *valid* in a given domain if it is true for every assignment of predicates and individuals from the given domain, and to be *satisfiable* in a given domain if it is true for some selection of predicates and individuals. A sentence valid in every domain is said to be universally valid, and a sentence satisfiable in some domain is said to be satisfiable. Thus either a sentence S is universally valid or its negation $\rightarrow$S is satisfiable. All provable sentences are universally valid and the converse is also true, so that predicate logic is also complete with respect to truth values.

Predicate logic is an example of an undecidable system, but various sub-systems of predicate logic are decidable. In particular monadic predicate logic, the logic of predicates with a single argument, is easily shown to be decidable.

Any predicate sentence (monadic or not) is decidable in a specified finite domain since in such a domain the concepts of 'all' and 'some' are equivalent simply to finite conjunctions and disjunctions. The special feature of monadic predicates is that a sentence containing k monadic predicate variables is universally valid if and only if it is valid in a domain with 2^k elements.

Let us suppose that some sentence S containing k monadic predicate variables is satisfiable in a domain D, containing more than 2^k individuals, by some choice $p_1, p_2, \ldots, p_k$ of predicates for the predicate variables $P_1, P_2, \ldots, P_k$ in S. We separate the elements of D into classes, two elements going into the same class only if the values of the predicates $p_1, p_2, \ldots, p_k$ are the same for the two elements. Since there at most 2^k different arrangements of truth values for k predicates, the elements of D fall into at most 2^k classes, $\alpha_1, \alpha_2, \alpha_3, \ldots, \alpha_n$ say, $n \leqslant 2^k$. Let e_r be an element of D in the class α_r, for $r = 1, 2, \ldots, n$ and for these values of r we define the predicate $q_i(r)$ to have the same truth value as $p_i(e_r)$, $i = 1, 2, \ldots, k$, in the domain F with elements $1, 2, 3, \ldots, n$; then by the definition of the q_i, sentence S is satisfiable in F by the predicates q_i if an only if S is satisfiable in D by p_i. The domain F may contain less than 2^k individuals but if S is satisfiable in F it will also be satisfiable in a larger domain with exactly 2^k individuals. By applying the foregoing discussion to $\rightarrow$S, we conclude that S is universally valid, if and only if it is valid in a domain of 2^k individuals.

Any predicate sentence may be expressed in so-called prenex normal form in which all quantifiers occur un-negated at the beginning and all operate on the same combination of predicates; this may be achieved by replacing each instance of a negated quantifier $\rightarrow AxP(x)$ and $\rightarrow ExP(x)$ by the equivalent forms $Ex\rightarrow P(x)$ and $(Ax)\rightarrow P(x)$ respectively, and then relettering the variables so that each quantifier has its proper variable and may be brought to the front of the sentence. The set of quantifiers in a prenex form is called the *prefix*, and the combination of predicates operated upon the *matrix* of the form.

Decision procedures are known for the universal validity of classes of sentences with a number of special prefixes. These comprise prefixes consisting solely of universal quantifiers or solely of particular quantifiers; prefixes in which all the universal precede all the existential quantifiers and prefixes with just one particular quantifier or two particular quantifiers not separated by a universal quantifier. For some other prefixes decision procedures are known only for special types of matrices, and for certain prefixes like $(Ex)(Ay)(Ez)$ and $(Ex)(Ay)(Ez)(Eu)$ the problem remains open.

I want to turn now from these purely logical instances of decidable systems to consider the decision problem for algebra. A particular fragment of algebra, called elementary algebra, consisting of the theory of polynomials of specified degree in any assigned number of real variables with integral coefficients is known to be decidable. The matrix of any sentence of elementary algebra in prenex form is a combination of equalities and inequalities between polynomials which may readily be brought into a disjunction of sentences of the form

$$\alpha_1 = 0 \ \& \ \alpha_2 = 0 \ \& \ \dots \ \& \ \alpha_m = 0 \ \& \ \beta_1 > 0 \ \& \ \beta_2 > 0 \ \& \ \dots \ \& \ \beta_n > 0$$

where each α and β is a polynomial of specified degree.

The decision procedure for elementary algebra is a method for eliminating quantifiers, transforming a sentence with quantifiers into an equivalent sentence without quantifiers or variables, the truth or falsehood of which may of course be determined simply by trivial calculations and sentence logic.

In essence the elimination procedure is a uniform method for determining the condition on the coefficients necessary and sufficient for the existence of a root of a set of polynomial equalities and inequalities. For instance the sentence

$$(Ay)(Ex)\{a_0 + a_1 x + a_2 x^2 + (b_0 + b_1 y + b_2 y^2)x^3 > 0\}$$

(where the a's and b's are integers) is equivalent to

$$(Ay)\{(a_0 > 0) \lor (a_1^2 > 4a_0 a_2) \lor (a_2 > 0) \lor \rightarrow (b_0 + b_1 y + b_2 y^2 = 0)\}$$

which in turn is equivalent to

$$(a_0 > 0) \lor (a_1^2 > 4a_0 a_2) \lor (a_2 > 0) \lor (4b_0 b_2 > b_1^2)$$
$$\lor \rightarrow \{(b_0 = 0) \ \& \ (b_1 = 0) \ \& \ (b_2 = 0)\}.$$

Elementary algebra is concerned solely with the fundamental field operations. Algebraic concepts whose definition involves set theoretic notions, or the notion of a natural number are excluded. Thus for instance although the sentence 'every cubic has a root' belongs to elementary algebra, the sentence 'every polynomial has a root' does not.

The exclusion of natural number variables is essential to the proof of decidability for, as we are now going to show, any system which contains natural number variables and both the operations of addition and multiplication is undecidable.

We start by considering a fragment of arithmetic, which we shall call system $\mathscr{A}$, based on predicate logic and the axioms

1. $Sx = Sy \rightarrow x = y$
2. $\rightharpoondown (0 = Sy)$
3. $\rightharpoondown (x = 0) \rightarrow Ey(x = Sy)$
4. $x + 0 = x, \quad x + Sy = S(x+y)$
5. $x \cdot 0 = 0, \quad x \cdot Sy = x \cdot y + x.$

The variables in these axioms are natural number variables, Sx symbolising the next number after x, so that the natural numbers themselves are represented in the system by 0, So, SSo, SSSo, and so on. We shall abbreviate the sign SSS . . . So with p S's to Np. The first axiom says that different numbers have different successors, and the second and third say that every number except o is the successor of some number. The groups of axioms 4 and 5 define addition and multiplication.

System $\mathscr{A}$ contains a certain amount of arithmetic but there are many simple theorems which are not provable in it. For instance none of

(i) $x \neq Sx, \quad 0 + x = x, \quad x + y = y + x, \quad 0 \cdot x = 0, \quad x \cdot y = y \cdot x$

is provable in $\mathscr{A}$. To show this we introduce a model which satisfies the axioms of $\mathscr{A}$ but not theorems (i). Such a model consists of the natural numbers 0, 1, 2, . . . combining in the usual way under addition and multiplication together with two extra numbers α and β such that

$S\alpha = \alpha, \quad S\beta = \beta, \quad \alpha + n = \alpha, \quad \beta + n = \beta, \quad n + \alpha = \beta, \quad n + \beta = \alpha,$
$\alpha \cdot 0 = 0, \quad \beta \cdot 0 = 0, \quad \alpha \cdot Sn = \beta, \quad \beta \cdot Sn = \alpha, \quad n \cdot \alpha = \alpha, \quad n \cdot \beta = \beta.$

It is readily verified that this model satisfies the axioms of $\mathscr{A}$ and therefore satisfies any sentence deduced from $\mathscr{A}$, but the model obviously does not satisfy theorems (i). What is lacking in $\mathscr{A}$ is mathematical induction, and if we added to the axioms of $\mathscr{A}$ all the infinitely many instances of the axiom

$$P(o) \ \& \ (Ax)\{P(x) \to P(Sx)\} \to AxP(x)$$

for all arithmetical predicates* P of the system (and an axiom making '$=$' a transitive relation) then we would be able to prove all the familiar theorems of classical number theory.

In spite of its deficiencies system $\mathscr{A}$ has a remarkable property; it can be shown that to every *computable* function $f(n)$, whose values may be computed on a machine with limited or unlimited memory, there corresponds a predicate $\phi(x, y)$ such that for any numbers N_x, N_y the sentence

$$\phi(N_x, N_y)$$

is provable or refutable in $\mathscr{A}$ according as y does or does not equal $f(x)$. The importance of system $\mathscr{A}$ lies in the fact that it attains this property with only a finite number of axioms.

We shall show that this fragment of arithmetic is undecidable; the proof of undecidability will not, however, be confined to system $\mathscr{A}$ but will apply to any consistent extension of $\mathscr{A}$ obtained by introducing additional axioms, for instance mathematical induction.

The proof of undecidability is based on Gödel's famous *arithmetisation of syntax*. Arithmetisation starts by assigning a number (which we take to be odd) to each arithmetical and logical sign. For instance we might be given the signs o, S, $=$, $+$, . the numbers 1, 9, 17, 25, 33, keeping other numbers of the form $8n + 1$ for brackets and logical signs; then individual variables might receive numbers of the form $8n + 3$ and predicate variables numbers of the form $8n + 5$.

A sentence made up of signs with the numbers

$$n_0, n_1, n_2, \ldots, n_k$$

(in this order) is given the number

$$2^{n_0} \cdot 3^{n_1} \cdot 5^{n_2} \ldots p_k{}^{n_k}$$

where p_k is the kth odd prime.

---

* An arithmetical predicate is a logical combination of equations between terms made up of variables and the signs o, S, $+$ and . ; e.g. $(Ex)(Ay)\{(x+y) \cdot z = Sz\}$.

Similarly a sequence of sentences (e.g. a proof) with numbers $s_1, s_2, \ldots, s_l$ is given the number

$$2^{s_0} \cdot 3^{s_1} \ldots p_l^{s_l}.$$

Given the number of a sentence, the sentence itself may be found by resolving the sentence number into its prime factors, and in the same way the constituent sentences of a proof may be regained from the number of the proof. Thus we have a code in which every symbol and every message is expressed by a single number.

It is comparatively simple to formulate the arithmetical relationship between the number of a conclusion and the numbers of the premises from which it is inferred by a given rule of inference, and hence to find the conditions which a number must satisfy to be the number of a proof. Arithmetisation, in fact, as the name suggests, maps the syntax of arithmetic on a part of arithmetic itself, and was used by Kurt Gödel in 1931 to prove that if arithmetic is consistent then this consistency is not provable in arithmetic.

In virtue of arithmetisation we may take the hypothesis that arithmetic is decidable to mean that there is a computable function Dn such that $Dn = 0$ if and only if n is the number of a valid sentence. If Dn exists there is a predicate $\Delta(x)$, say, in $\mathscr{A}$, such that either $\Delta(N_n)$ or $\to\Delta(N_n)$ is valid according as Dn is or is not zero.

Denote by Sub n the number of the sentence obtained by substituting the numeral N_n for the free variable in sentence number n; Sub n is obviously computable and so there is a predicate $\Sigma(x, y)$, say, such that

$$\Sigma(N_m, N_n)$$

is valid if and only if $n = \text{Sub } m$.

Let **p** be the number of the sentence

$$(Ay)\{\Sigma(x, y) \to \to\Delta(y)\}$$

so that Sub **p** is the number of the sentence

$$(Ay)\{\Sigma(N_{\mathbf{p}}, y) \to \to\Delta(y)\},$$

which we denote by P.

If P is valid in $\mathscr{A}$ then

$$\Sigma(N_{\mathbf{p}}, N_{\text{Sub } \mathbf{p}}) \to \to\Delta(N_{\text{Sub } \mathbf{p}})$$

is valid; since the antecedent $\Sigma(N_p, N_{Sub\ p})$ is valid, by definition, it follows that if P is valid then $\rightarrow\!\Delta(N_{Sub\ p})$ is valid. On the other hand, if P is not valid, then $\rightarrow\!\Delta(N_{Sub\ p})$ is valid, since Sub **p** is the number of P, and so

$$\rightarrow\!\Delta(N_{Sub\ p})$$

is valid in either case, from which it follows that

$$\Sigma(N_p, N_{Sub\ p}) \rightarrow \ \rightarrow\!\Delta(N_{Sub\ p})$$

is valid.

But if $y \neq N_{Sub\ p}$ then $\Sigma(N_p, y)$ is false and therefore

$$\Sigma(N_p, y) \rightarrow \ \rightarrow\!\Delta(y)$$

is valid.

It follows that

$$(Ay)\{\Sigma(N_p, y) \rightarrow \ \rightarrow\!\Delta(y)\}$$

is valid, i.e. sentence number Sub **p** is valid and so

$$\Delta(N_{Sub\ p})$$

is valid and arithmetic contains a contradiction.

Thus if arithmetic is free from contradiction it is undecidable.

The undecidability of predicate logic is deducible from the undecidability of system $\mathscr{A}$. A necessary preliminary to the derivation is the elimination of the extra-logical predicate and individual symbols and the operators $+$ and $.$ from $\mathscr{A}$. To this end we write the predicate $E(x, y)$ for the equation $x = y$, the predicate $S_0(x, y)$ for the equation $Sx = y$, and the predicates $S_1(x, y, z)$, $P_1(x, y, z)$ for the equations $x + y = z$ and $x \,.\, y = z$ respectively. To eliminate the constant o we use axioms 2 and 3 and write $(Az) \rightarrow (x = Sz)$ for the equation $x = o$. Under these transformations axioms 2 and 3 become universally valid, axiom 1 becomes

$$S_0(x, z) \ \& \ S_0(y, z) \rightarrow E(x, y),$$

and the axioms for addition and multiplication are transformed into

$$\{(Az) \rightarrow S_0(z, w)\} \rightarrow S_1(x, w, x)$$
$$\{S_0(y, u) \ \& \ S_0(v, w) \ \& \ S_1(x, y, v)\} \rightarrow S_1(x, u, w)$$

and
$$\{(Az) \rightarrow S_0(z, w)\} \rightarrow P_1(x, w, w)$$
$$S_0(y, u) \ \& \ S_1(v, x, w) \ \& \ P_1(x, y, v) \rightarrow P_1(x, u, w).$$

Let the conjunction of these transforms (of which there are only a finite number) be denoted by F, and consider any sentence $S_{\mathscr{A}}$ in system $\mathscr{A}$. By the given transformations $S_{\mathscr{A}}$ becomes a predicate sentence S say, and $S_{\mathscr{A}}$ is provable in $\mathscr{A}$ if, and only if, S is derivable from F in predicate logic. But by the deduction theorem, S is derivable from F if, and only if,

$$F \to S$$

is provable in predicate logic, regarding E, S_0, S_1 and P_1 as predicate variables. Hence if predicate logic were decidable then $F \to S$ would be decidable; if $F \to S$ is universally valid then since F is provable in $\mathscr{A}$, $S_{\mathscr{A}}$ is provable in $\mathscr{A}$, and if $\to (F \to S)$, i.e. F & $\to S$ is satisfiable then $S_{\mathscr{A}}$ is not provable in $\mathscr{A}$. Thus if predicate logic were decidable we should have a decision procedure for the undecidable system $\mathscr{A}$.

BIBLIOGRAPHICAL NOTES

(The numbers in square brackets refer to the Bibliography)

The Arithmetisation of Syntax was introduced by K. Gödel in [2].

The first proof of the undecidability of predicate logic was given by A. Church in [1].

A decision method for monadic predicate logic was first given by L. Löwenheim in [3].

The truth-table decision method for sentence logic was first obtained by L. Wittgenstein in [9] and E. Post in [4].

A decision method for Arithmetic without multiplication was given by M. Presburger in [5].

The foregoing proof of the essential undecidability of Arithmetic is due to A. Tarski, A. Mostowski and R. M. Robinson and is given in [8].

The decision method for elementary algebra was obtained by A. Tarski and written up by J. C. C. McKinsey in [7], and that for Abelian Groups is given by Wanda Szmielew in [6].

REFERENCES

1. Church, Alonzo. 'A note on the Entscheidungsproblem', *Journal of Symbolic Logic*, Vol. I (1936), pp. 40-1, 101-2; 'An unsolvable problem of elementary number theory', *American Journal of Mathematics*, Vol. 58 (1936), pp. 345-63.

2. Gödel, Kurt. 'Über formal unentscheidbare Satze der Principia Mathematica und verwandter Systeme I', *Monatshefte für Mathematik und Physik*, Vol. 38 (1931), pp. 173-98.
3. Löwenheim, Leopold. 'Über Möglichkeiten in Relativkalkül', *Mathematische Annalen*, Vol. 76 (1915), pp. 447-70.
4. Post, Emil L. 'Introduction to a general theory of elementary propositions', *American Journal of Mathematics*, Vol. 43 (1921), pp. 163-85.
5. Presburger, M. 'Über die Vollständichkeit eines gewissen Systems der Arithmetik ganzer Zahlen, in welchem die Addition als einzige Operation hervortritt', *Comptes-rendus du I Congrès des Mathématiciens des Pays Slaves* (Warszawa, 1929), pp. 92-101 and 395.
6. Szmielew, Wanda. 'Decision problem in group theory', *Proceeding Xth International Congress of Philosophy* (Amsterdam, 1948), fasc. 2, pp. 763-6.
7. Tarski, Alfred. *A decision method for elementary algebra and geometry.* California, 1948.
8. Tarski, Alfred; Mostowski, Andrzej; and Robinson, Raphael M. *Undecidable Theories.* Amsterdam, 1953.
9. Wittgenstein, Ludwig. *Tractatus logico-philosophicus.* London, 1922.

XI

THE SIGNIFICANCE OF INCOMPLETENESS
THEOREMS

I WANT to start by considering certain fundamental differ-- ences between the major incompleteness theorems which have been discovered in researches in the foundations of mathematics during the past thirty years and the incomplete axiom systems which were found in the study of projective geometry during the last century. A confusion between these forms of incompleteness has led some mathematicians to under- estimate the significance of the newer results and some philo- sophers to seek to understand the meaning of the new discoveries by reference to the technically simpler older work.

When it was discovered, for instance, that a system of pro- jective geometry in two dimensions which postulated the axioms of incidence was incomplete because Desargues' Theorem on perspective triangles was not derivable in the system, then this incompleteness could be interpreted as a proof of the indepen- dence of Desargues' Theorem, postulated as a new axiom, from the other axioms. The impossibility of proving Desargues' Theorem is *surprising* in view of the fact that the corresponding theorem in three dimensions is derivable from the three- dimensional axioms of incidence, but it is nevertheless, I think, without philosophical significance because it throws no light on the nature of formal systems as such and imposes no limitations upon the axiomatic method.

The great modern incompleteness theorems which I shall consider are those due to Skolem and Gödel. Skolem's incom- pleteness theorem was discovered in an attempt to explain a paradox which Skolem himself found in the theory of sets.

The paradox out of which Skolem's incompleteness theorem arises, is produced by applying a result of Löwenheim's to a formalised set theory. Löwenheim showed that every consistent set of statements has a denumerable model, and so any formal system which admits some model (of the power of the continuum perhaps) has also a *denumerable* model. That is to say, for a consistent theory, we can find an interpretation in which all the objects, of which the theory treats, may be taken to be the natural numbers. Consider now some formalisation of set theory; according to the Löwenheim theorem we can find an interpretation of the membership relation of the theory in which all the *sets* of the theory are taken to be natural numbers. But in any adequate formalisation of set theory, using the familiar diagonal process, we can prove Cantor's theorem that the set of all subsets has a *greater* cardinal than the set itself:

Let S be a denumerable set and let a subset of S be denoted by a sequence of zeros and units, a zero in the nth place showing that the nth member of S is *not* in the subset, and a unit in the nth place showing that the nth member of S is a member of the subset. Suppose now that the set of all subsets of S *is* denumerable, and let it be enumerated as follows:

$$s_1 = a^1_1,\ a^1_2,\ a^1_3,\ \ldots$$
$$s_2 = a^2_1,\ a^2_2,\ a^2_3,\ \ldots$$
$$s_3 = a^3_1,\ a^3_2,\ a^3_3,\ \ldots$$

where each a^n_r is either 0 or 1.

Define $$b_n = a^n_n + 1 \quad (\mathrm{mod}\ 2)$$

and consider the subset

$$\sigma = b_1,\ b_2,\ b_3,\ \ldots$$

The subset σ differs from S_1 in respect of the first element of S since b_1 is 1 or 0 according as a^1_1 is 0 or 1, from S_2 in respect of the second element, and so on. Thus the subset σ does *not* occur in the enumeration $S_1, S_2, S_3, \ldots$ and the hypothesis that the set of all subsets of S is enumerable is disproved.

But by the Löwenheim theorem there is a model of set theory (supposed consistent) in which each set is associated with a natural number, so that in defiance of Cantor's theorem, the set of subsets *is denumerable*.

This is Skolem's paradox. The conclusion which Skolem himself drew from the paradox is that a formalisation of set theory can contain only *relatively non-denumerable* sets; i.e. sets which are non-denumerable only because the formalisation lacks the functions to enumerate them. In other words every formalisation of set theory must be incomplete in the sense that there are denumerable sets which cannot be *proved* denumerable within the theory. To justify this interpretation of the paradox one must observe that the proof of Cantor's theorem in some system starts by assuming that a certain mapping of a set on its subsets *exists* and derives a contradiction *in the system* from this assumption. Existence here of course means existence in the system, so that the conclusion to be drawn from the contradiction is, not that the mapping in question does not exist, but only that it does not exist *in the system*. Since Löwenheim's theorem assures us that the mapping does in fact exist, it follows that *there is a mapping which is not contained in the formal system*, so that the system is incomplete with respect to the class of mappings it contains. Unlike the situation in projective geometry we cannot remedy the deficiency by fortifying the system with another axiom; we could complete the system only at the price of rendering it inconsistent. Another way of expressing the result is to say that Löwenheim's theorem for any particular formalisation of set theory is not provable by means of the resources of that formalisation alone.

Of course the notion that there are no absolutely non-denumerable sets is not a new one. The sole ground we have for believing in the existence of a non-denumerable set lies in Cantor's theorem itself. But if we do *not* assume that the totality of subsets forms a set (and this is nothing *but* an assumption) then all that the diagonal process proves is that from *any* sequence of subsets we can construct another subset, just as from any natural number we can construct another, by adding one. And if we give up the axiom of subsets of course the Skolem paradox disappears.

There is nothing in the paradox itself to *force* us to give up the axiom of subsets. A constructivist who already rejects the set of all subsets on other grounds will not need to reckon with the Skolem paradox; and a mathematician seeking the greatest possible generality will have to remain content with *relative* non-denumerability. At best we can have a transfinite hierarchy of systems in each of which there are sets non-denumerable in a particular system but denumerable in a system of greater ordinal.

The Gödel Incompleteness Theorem

I come now to the major incompleteness theorem of mathematical logic, Gödel's Theorem that all sufficiently rich formal systems necessarily contain sentences which are neither provable nor refutable in the system. There are so many interesting facets to this result that I shall later consider the proof in some detail, but first I want to observe that in this theorem, as in the Skolem theorem for sets, the incompleteness revealed by the theorem cannot be filled by means of a new axiom; it is true that the particular undecidable sentence constructed in the proof can itself be postulated as an axiom, but the proof shows that the system so fortified will still contain undecidable sentences. The undecidable sentence in this theorem of Gödel has the form $(\forall x)P(x)$; neither $(\forall x)P(x)$ itself, nor its negation $(Ex)\rightarrow P(x)$ is provable in the system under consideration, system $\mathscr{F}$ say, but each instance of the universal sentence $P(x)$, namely $P(o)$, $P(1)$, $P(2)$, ... *is* provable in $\mathscr{F}$. Many attempts have been made to close the gap this theorem is thought to reveal in the proof structure of formal systems. The most obvious course to take would be to add to the proof resources of the system a new derivation scheme permitting the derivation of $(\forall x)P(x)$ from the infinite sequence of sentences

$$P(o), P(1), P(2), \ldots;$$

but this device entirely destroys the finite character of a proof process. In the Gödel Theorem as we shall see, the very possibility of *proving* an infinite number of sentences $P(o)$, $P(1)$, ... without a prior proof of the universal sentence $P(x)$, was revealed for the first time. This aspect of Gödel's construction has been cleverly exploited in a recent attempt to obtain a (relatively) closed proof system without introducing non-finitist proof schemata, and is in some respects one of the most interesting features of the Gödel Theorem.

As is well known, the heart of Gödel's construction is a one-to-one mapping of the syntax of a formalised arithmetic upon arithmetic itself. There are many ways known in which this mapping may be accomplished but I shall simply suppose that each primitive sign of some formal system of arithmetic $\mathscr{A}$ has been assigned a number, and that each sequence of primitive signs with numbers $n_0, n_1, n_2, \ldots, n_k$, whether forming a sentence or not, is given the number $2^{n_0} \cdot 3^{n_1} \ldots \cdot p_k^{n_k}$, where p_k is the kth odd prime number. I shall further suppose that the formal system $\mathscr{A}$ contains all primitive recursive functions, either directly in the sense of admitting primitive recursive definitions as axioms or indirectly by having definition resources like existential and minimal operators. I may mention in passing that it is the failure to introduce this requirement and to show the fundamental part it plays, which vitiates most popular accounts of Gödel's work. I shall also assume that $\mathscr{A}$ is recursively axiomatisable, so that the predicate 'n is the number of an axiom' is expressed by a primitive recursive relation $A(n)$, i.e. is expressed within $\mathscr{A}$ itself by this relation. Such relations as 'n is the number of a one variable primitive recursive function', 'n is the number of a variable in formula number f', 'n is the number of a proof of formula f' and 'n is the number of the formula which results by substituting the numeral representation of number k in formula f' may all be shown to be primitive recursive. The key tools in establishing these results are the primitive recursiveness of the relation

$$(Ex)\{x \leqslant y \ \& \ R(x, y, z)\}$$

where R itself is primitive recursive, and the reductions to primitive recursive definition of a related schema of definition known as definition by course-of-values recursion. To exhibit Gödel's undecidable sentence I denote by

$$St_f(\nu/r)$$

the number of the expression of the formal system $\mathscr{A}$ obtained by substituting the numeral representing the number r for the variable of number ν in the formula with number f, and by

$$Pr(m, n)$$

the relation which says that m is the number of a proof of formula number n. Further let ν be the number of the variable n and let $\mathbf{a}$ be the representation of the number of the sentence of $\mathscr{A}$ which we are denoting by

$$(\forall m) \rightarrow \Pr(m, St_n(\nu/n)) \qquad \ldots \text{(i)}$$

and finally let $(\forall m)G(m)$ denote the sentence obtained from (i) by substituting the numeral **a** for the variable n. Then neither $(\forall m)G(m)$ nor its negation is provable in $\mathscr{A}$. We observe first that since $(\forall m)G(m)$ is formed by substituting numeral **a** for the variable n in formula number **a**, its number is therefore

$$St_a(\nu/a).$$

Hence if $(\forall m)G(m)$ were provable, and if **k** were the number of its proof, then

$$\Pr(\mathbf{k}, St_a(\nu/a)) \qquad \ldots \text{(ii)}$$

holds (and the formula in $\mathscr{A}$ which this represents is provable in $\mathscr{A}$, *because* $\Pr(m, n)$ is a primitive recursive relation, and this is one of the points where this fact is critical to the proof); but this contradicts the formula which $(\forall m)G(m)$ itself represents, i.e. that represented by

$$(\forall m) \rightarrow \Pr(m, St_a(\nu/a)).$$

I emphasise again that the contradiction is in $\mathscr{A}$ itself, and so if $\mathscr{A}$ is consistent then $(\forall m)G(m)$ is not provable in $\mathscr{A}$. To prove that $\rightarrow (\forall m)G(m)$ is also unprovable I shall assume rather more than the consistency of $\mathscr{A}$, the so-called ω-consistency of $\mathscr{A}$, but this additional assumption could be dispensed with at the price of taking a rather more complicated sentence than $(\forall m)G(m)$. By the ω-consistency of $\mathscr{A}$ we mean that for any formula $\mathscr{G}(m)$ it is impossible to prove in $\mathscr{A}$ that $(Em) \rightarrow \mathscr{G}(m)$ and $\mathscr{G}(o)$, $\mathscr{G}(1)$, $\mathscr{G}(2)$ ... all hold simultaneously. First we observe that as a consequence of what we have already established, if $\mathscr{A}$ is simply consistent, then none of the numbers $0, 1, 2, \ldots$ is the number of a proof of formula number $St_a(\nu/a)$, and so

$$\rightarrow \Pr(m, St_a(\nu/a))$$

is provable in $\mathscr{A}$ for any m, i.e. $G(o), G(1), G(2), \ldots$ are all provable. Hence by ω-consistency, $(Em) \rightarrow G(m)$ is *not* provable. Thus we have seen that neither $(\forall m)(Gm)$ nor its contrary is provable in $\mathscr{A}$, although each instance of the general formula, viz. $G(o), G(1), G(2), \ldots$ is provable. The formula $(\forall m)G(m)$ is said to be undecidable.

I remarked earlier that Gödel's arithmetisation showed for the first time how it is possible in a formal system with finite proof procedure to prove all the formulae G(o), G(1), . . . *without first proving* G(m), (in fact even if G(m) is not provable); we have just seen an instance of this. In the general case, let N(k), be the numeral representing k in A, and let **h** be the number of a formula H(n), then the assertion that *all* the formulae H(o), H(1), . . . are provable is expressed by

$$(\forall k)(\mathrm{E}m)\mathrm{Pr}(m, \mathrm{S}t_h(v/\mathrm{N}(k))$$

and this formula may be provable in $\mathscr{A}$ even though H(n) is not. This constitutes a *formalisation of the notion of an arbitrarily assigned integer*.

The specific instance of an undecidable formula $(\forall m)$ G(m) which is constructed in Gödel's proof is of no particular significance in arithmetic, but by formalising the proof of undecidability Gödel obtained the remarkable conclusion that the sentence of arithmetic which entails arithmetic's freedom from contradiction, viz.

$$(\forall m)(\forall r)(\forall s)\rightarrow\{\mathrm{P}(r, m)\ \&\ \mathrm{P}(s, \mathrm{Neg}\ m)\} \qquad (\mathrm{C})$$

(where Neg m is the number of the negation of sentence number m) is itself undecidable, if arithmetic is consistent. Contrary to a widely held belief this result does not, however, establish the impossibility of proving the consistency of a codification of arithmetic by finitist methods formalisable within the codification. Even though the closed formula C is not provable in $\mathscr{A}$, each of its instances

$$\rightarrow\mathrm{P}(\mathbf{r}, \mathbf{m})\ \&\ \mathrm{P}(\mathbf{s}, \mathrm{Neg}\ \mathbf{m})$$

is provable in $\mathscr{A}$ for arbitrary **r, s, m** (in virtue of Gentzen's consistency proof by transfinite induction), and the general formula expressing the provability of these instances may itself be provable in $\mathscr{A}$. But of course a proof inside $\mathscr{A}$ of $\mathscr{A}$'s consistency offers no security, for if $\mathscr{A}$ were inconsistent then every formula in $\mathscr{A}$ would be provable in $\mathscr{A}$. In fact, Gödel's result does not really bear upon the problem of consistency itself but affords a means of establishing the independence from the axioms of $\mathscr{A}$ of axioms (like transfinite induction) whose addition to $\mathscr{A}$ suffice to prove the closed formula C above in the enlarged system.

If we add the unprovable formula $(\forall m)G(m)$ to $\mathscr{A}$ as a new axiom forming a system A^+ say, then exactly as before we can construct an undecidable formula in A^+, so that A^+ is also incomplete; even if we form a new system B, by adding in all the undecidable formulae of $\mathscr{A}$ as additional axioms, if the axioms of B form a recursive set, then B is still incomplete. Since $(\forall m)\ G(m)$ is not provable in $\mathscr{A}$ the system A^- formed by adding the *denial* of $(\forall m)G(m)$ is consistent if $\mathscr{A}$ is consistent, and so by Löwenheim's Theorem A^- must admit a denumerable model in which $G(o)$, $G(1)$, ... are all true but $(\forall m)G(m)$ is false and therefore m must take values other than o, 1, 2, ...

It follows from this that A^- must admit what is called a non-standard model, that is, an interpretation in which a class of objects which is not ordinally similar to the natural numbers plays the part of the natural numbers. That this in fact is the case was shown independently by Skolem in 1934, who proved that a certain class of functions can play the number role.

The construction of the formula $G(m)$ above can also be carried out in a system without quantifiers, in some formalisation of recursive arithmetic, $\mathscr{R}$ say, and we find that $G(m)$ with free variable m is unprovable in $\mathscr{R}$, but each of $G(o)$, $G(1)$, ... is provable. We cannot therefore explain away Gödel's incompleteness theorem as a defect of quantification theory. $\mathscr{R}$ like $\mathscr{A}$ is also incomplete. There is however an important difference between $\mathscr{R}$ and $\mathscr{A}$ since it can be shown that no non-standard model of $\mathscr{R}$ can itself be a recursive model.

Discussing Gödel's incompleteness theorem in 1934, before Skolem's result was known here, Wittgenstein was led to the same interpretation of the theorem, that induction and substitution of natural numbers for free variables fail to ensure that the natural numbers are the *only* values which the variables may take; it is perhaps surprising that the passages on Gödel's Theorem in the recently published 'Reflections on the Foundations of Mathematics' give no hint of this remarkable insight.

Since every axiom system for the natural numbers is incomplete and therefore necessarily admits a non-standard model it has been argued that we must look outside axiom systems for a logical foundation of arithmetic; this view is associated with a neo-realist outlook in foundation studies, that the elements of mathematics are objects in a real world with so-to-speak physical properties which the mathematician only partially captures in an axiom system. Certainly in some of his work the mathematician has an almost overwhelming conviction that he is uncovering connections which lie waiting to be revealed, but this may only reflect his amazement at the astonishing way the pieces of a puzzle sometimes fit together; we feel that the pieces must have been made to fit before we actually handled them, that we are only reconstructing a puzzle some other mind set for us. Neo-realism is not, however, a return to the Greek standpoint that formal geometry is an account of the space of our physical sensations. The case against classical philosophical realism in mathematics is overwhelmingly strong. In geometry the well-known consistency proofs of non-Euclidean geometry relative to the Euclidean makes it impossible for only one of the two geometries to be valid and yet both cannot mirror the real world. The neo-realist argues from this, not that the elements of geometry are *concepts*, but that no formal system can adequately express the whole of *geometry*, which is something revealed only to the intuition. This is an attractive thesis; every mathematician is conscious of possessing an 'inner sight', an inward short cut. But as a philosophical analysis of mathematics, neo-realism is no more tenable a position than classical realism. Intuition can be false and misleading and the inward 'short cut' just a *cul de sac*; intuition is certainly an important element in the creation of mathematics, but to see it as the organ which gives the mathematician access to mathematical reality is to be deceived by an analogy. When we say that no formal system can characterise the number concept, we do not mean that the number concept is something which we already have independently of the formal system; I may reject every definition of the meaning of a word, because it fails to characterise what I mean by the word, and maintain, rightly, that I know well what the meaning is, and yet my knowing what the meaning is may consist in nothing more than my rejection of the definitions. Just as I may write a story and be left with the feeling that this is not the story I meant to write, although of course I have not already in mind another story with which I compare it. When we contrast formal mathe-matics with intuitive mathematics we are not contrasting an image with reality, but a game played according to strict rules with a game with rules which change with the changing situa-tion; a proof in intuitive mathematics may be a particular way of looking at a diagram, i.e. a particular way of using it as a symbol, or it may consist in stepping outside the particular system with which we are operating.

What we call an intuitive proof of some particular mathe-matical relation, is not a proof intelligible only to some special sense, quite the contrary. An intuitive proof of the relation is a proof which makes the minimum appeal to esoteric knowledge, which links the relation most immediately to a familiar back-ground; but in its role as proof an intuitive proof has the same essential character as a formal proof, it exposes the connections between one relation and another.

An intuitive proof may for instance be a proof in which generality is expressed without the use of variables. For instance I may prove the general theorem $ab = ba$ without introducing variables, by looking at the array

$$
\begin{matrix}
\cdot & \cdot & \cdot & \cdot & \cdot & \cdot & \cdot \\
\cdot & \cdot & \cdot & \cdot & \cdot & \cdot & \cdot \\
\cdot & \cdot & \cdot & \cdot & \cdot & \cdot & \cdot \\
\cdot & \cdot & \cdot & \cdot & \cdot & \cdot & \cdot \\
\cdot & \cdot & \cdot & \cdot & \cdot & \cdot & \cdot
\end{matrix}
$$

first as five rows of seven dots and then as seven columns of five dots. What is *general* now is the *method* of proof; to show some one that the proof *is* general, it may be necessary to write it out again, with different numbers of dots, but this only means that we seek to draw attention to certain features of the proof, not that the proof appeals to a different *sense* than a proof which uses symbols for generality. The proof is just as formal as the proof with variables and quantifiers. What a single formal system is unable to do is to comprehend all possible partial systems in a single whole—only in this sense are formal systems necessarily incomplete; the only 'reality' with which we can contrast a formal system is another system in a hierarchy of more or less uniformly formalised systems.

There is another incompleteness theorem for $\mathscr{A}$ which has no parallel in $\mathscr{R}$. It can be shown (for instance by means of the well-known result of classical analysis that a bounded monotonic increasing sequence is convergent) that there is a formula $F(x)$ in $\mathscr{A}$ such that $(Ex)F(x)$ is provable in $\mathscr{A}$, but none of $F(o)$, $F(1)$, $F(2)$, . . . is provable. The formula F is this incompleteness theorem itself contains a quantifier, and no example has yet been found of a primitive recursive predicate R such that $(Ex)\,R(x)$ is provable (in some formalisation of arithmetic) and yet none of $R(o)$, $R(1)$, $R(2)$, . . . is provable.

Gödel's construction of an undecidable sentence in a formal system $\mathscr{A}$ utilises a detailed knowledge of the proof procedure of the system and the undecidability may be thought to reflect this structure. Kleene has however given a uniform process for finding an undecidable formula in every suitable consistent formal system. By means of Gödel arithmetisation one may determine a primitive recursive relation $T(z, x, y)$ such that for $z = 0$, 1, 2, . . . ,

$$(Ey)T(z, x, y)$$

enumerates (with repetitions) all relations of the form $(Ey)R(x, y)$ with general recursive R. Hence given any general recursive R we may determine r so that

$$(Ey)R(x, y) = (Ey)T(r, x, y).$$

Let S be a formal system in which every general recursive function may be expressed and evaluated and such that for any formula $F(x)$ of S there is a general recursive relation $R_F(x, y)$ which holds only when y is the number of a proof of $F(x)$ in S.

Let $t(x, y)$ be the representation in S of the primitive recursive relation $T(x, x, y)$, let $A(x)$ stand for $(\forall y) \rightarrow t(x, y)$, and let r be the number given above such that

$$(Ey)R_A(x, y) = (Ey)T(r, x, y) \qquad \ldots (3)$$

Then if S is consistent and ω-consistent, neither $A(r)$ nor $\rightarrow A(r)$ is provable in S.

For if $A(r)$ is provable, let p be the number of its proof, so that $R_A(r, p)$ holds, whence by (3) there is an η such that $T(r, r, \eta)$ holds, and therefore $t(r, \eta)$ is provable; but if $A(r)$ is provable, $\rightarrow t(r, \eta)$ is provable. This contradiction in S shows that $A(r)$ is not provable in S, and therefore

$$\rightarrow(Ey)R_A(r, y) \equiv \forall(y) \rightarrow T(r, r, y)$$

holds; consequently $\rightarrow t(r, y)$ is provable for each y, and therefore $(Ey)t(r, y)$ is unprovable, i.e. $\rightarrow A(r)$ is unprovable.

Kleene's procedure can be applied only to systems with quantifiers and in this respect is less general than Gödel's original construction which is applicable also to a free variable system. It is perhaps also worth noting that even in Kleene's procedure the actual instance of an undecidable sentence is a function of the formal system being considered, since the constant r in $A(r)$ is determined by the proof predicate $R_A(xy)$ and this of course varies from system to system. But the most interesting feature of the proof is the contrast of the semi-formal predicate $T(x, x, y)$ with its intended formal counterpart $t(x, y)$. In the theory of the predicate $T(x, x, y)$ we suppose we have before us a system of equations from which the value of a function $f(x)$ is derived by repeated substitution, a certain incompletely defined auxiliary function in the equations yielding a value only when we reach a value of y such that $R(x, y)$ holds; we then use Gödel numbering to replace the syntactical notion of deriving from a system of equations by a primitive recursive function (which we may for present purposes identify with) $T(r, x, y)$, r being the number of the system of equations and y the number of the derivation of an end equation $f(x) = \zeta$ from the system; the final step is to link $R(x, y)$ and $T(r, x, y)$ by the observation that there is a derivation of an end equation if and only if there is a y for which $R(x, y)$ holds. We are not concerned here with a formula in a formal system

$$(Ey)R(x, y) = (Ey)T(r, x, y)$$

but, as we say, with an assertion of *existence*. What does this mean? In this case that we have a procedure that enables us to *construct* the y for which $T(r, x, y)$ holds from a knowledge of the y for which $R(x, y)$ holds and conversely. But of course this *procedure* is a purely formal procedure. The situation is exactly akin to a familiar application of mathematics. If oranges cost 3d. each then I must pay 1/3d. for five; the reason for paying 1/3d. is the formal equation

$$5 \times 3 = 12 + 3,$$

but the actual purchase and payment lie outside the formal system. Depicting a computation procedure in this way (without necessarily using it) is one of the things we mean by an intuitive proof.

It is often said that Gödel's formula $(\forall m)G(m)$ is *true* but unprovable. The reason for saying that it is true is presumably that since each of $G(0)$, $G(1)$, $G(2)$, . . . is provable, and so true, therefore $G(m)$ is true for all m, which is just another way of saying that $(\forall m)G(m)$ is true. Of course if we do mean nothing more by saying that $(\forall m)G(m)$ is true than that $G(m)$ is true for all m then it is certainly true to say that $(\forall m)G(m)$ is true but unprovable. But the expression is a rather misleading one. The relationship between the formal system and the metalanguage which is established by recursion assures us that if $R(m)$ is a primitive recursive predicate such that $R(\mathbf{m})$ holds for some $\mathbf{m}$ then certainly $R(\mathbf{m})$ is provable in $\mathscr{A}$, or rather the formula in $\mathscr{A}$ which represents $R(\mathbf{m})$ is provable. But it is the essence of Gödel's theorem itself that although $G(m)$ is primitive recursive the formula $(\forall m)G(m)$ does *not* express the notion 'for all m, $G(m)$' in the formal system. As we have seen there is an interpretation of the system in which $G(0)$, $G(1)$, $G(2)$, . . . are *not* all the instances of $(\forall m)G(m)$ and therefore the truth of these instances is not to be identified with the truth of the formula $(\forall m)G(m)$.

Another common mistake is to suppose that $(\forall m)G(m)$ is true because it truly affirms of itself that it is non-demonstrable. We recall that $G(m)$ is an abbreviation for the formula which is obtained by substituting the number $\mathbf{a}$ of the formula

$$(\forall m)\rightarrow\mathrm{Pr}(m, \mathrm{St}_n(\mathbf{v}/n))$$

for the variable n in this formula.

The number of the resulting formula is of course

$$\mathrm{St}_\mathbf{a}(\mathbf{v}/\mathbf{a})$$

as we already have had occasion to remark; hence the fact that $(\forall m)G(m)$ is unprovable, i.e. that formula number $\mathrm{St}_\mathbf{a}(\mathbf{v}/\mathbf{a})$ is unprovable, tells us that

$$\rightarrow\mathrm{Pr}(m, \mathrm{St}_\mathbf{a}(\mathbf{v}/\mathbf{a}))$$

is provable for each value of m, but this of course tells us nothing about the formula

$$(\forall m) \rightarrow \Pr(m, St_a(v/a))$$

that is, nothing about $(\forall m)G(m)$. The Gödel numbering establishes a code in which each instance of the numerical formula $G(\mathbf{m})$ says that $\mathbf{m}$ is not the number of the proof of $(\forall m)G(m)$, but $(\forall m)G(m)$ itself says nothing at all in the code. Thus the Gödel sentence is neither an example of self-reference nor of self-description. Even the sense in which we can say that the formula of $\mathscr{A}$ which we are denoting by

$$\Pr(\mathbf{x}, \mathbf{y}) \tag{P}$$

says that $\mathbf{x}$ is the number of the proof of formula number $\mathbf{y}$ is in need of clarification. As a formula of $\mathscr{A}$, P says nothing at all. As the representative in $\mathscr{A}$ of a certain arithmetical relation it says that y is the exponent of the greatest power of the greatest prime number which divides x; and only as a sentence of the code which the Gödel numbering establishes, does this arithmetical relation say that $\mathbf{x}$ is the number of the proof of formula number $\mathbf{y}$.

Even supposing, which is not in fact the case, that there is a formula of the formal system (let us call it ϕ) such that *as a sentence of the code* ϕ says something about the formula ϕ of the formal system, we still could not claim that ϕ is an example of successful self-reference or self-description, for as an element of the formal system ϕ is just a sign pattern, and as a sentence of the code ϕ refers not to itself i.e. not to its meaning, but to the sign by which it is expressed, in the way the sentence

'This is written in chalk'

refers to its physical character, not to its sense.

What a sentence affirms depends upon the language in which the sentence is being used as a sentence, and identical sentences may express different propositions in different languages, as the obvious example of a code in which every English sentence stands for its contrary shows. But we must not, therefore, suppose, as many logicians do, by analogy with the specification of the range of a variable in mathematics, that every sentence p

must be qualified by another sentence which names the language to which p belongs; for this assumption leads to an infinite hierarchy of languages, without achieving its aim. Whether we are speaking a common language or not cannot ultimately be settled by language alone but must show itself in our actions.

INDEX